CW00631349

S247 INORGANIC CHEMISTRY
CONCEPTS AND CASE STUDIES
SCIENCE: A SECOND LEVEL COURSE

BLOCK 7
THE CHEMISTRY OF GROUPS III–VI

PREPARED FOR THE COURSE TEAM BY
DAVID JOHNSON AND LESLEY SMART

THE OPEN UNIVERSITY

S247 REWRITE COURSE TEAM

CHAIR
David Johnson (1989–90)
Lesley E. Smart (1990–94)

AUTHORS
David Johnson
Elaine A. Moore
Jane Nelson
Lesley E. Smart

CONSULTANT
Peter Timms, *University of Bristol* (Block 7)

COURSE COORDINATOR
Wendy L. Selina

TEXT PROCESSING SERVICES
Pam Berry

EDITORS
Ian Nuttall
Dick Sharp

BBC
Barrie Whatley
Gail Block

DESIGN GROUP
Debbie Crouch (designer)
Alison George (graphic artist)

SECRETARIAL SUPPORT
Jenny Burrage
Sally Eaton
Sue Hegarty

ORIGINAL S247 COURSE TEAM

CHAIR
Stuart Bennett

AUTHORS
Stuart Bennett (Blocks 2 and 3; Case Study 4)
Charles Harding (Block 6)
David Johnson (Block 1; Case Study 1)
Joan Mason (Block 5)
Elaine A. Moore (Block 4)
Jane Nelson (Block 7)
Lesley E. Smart (Blocks 2 and 7)
Keith Trigwell (Case Study 2)
Kiki Warr (Block 1)
John Emsley (Block 7; Case Study 4)
Harry Kroto (Case Study 3)

COURSE COORDINATOR
Keith Trigwell

The Open University, Walton Hall, Milton Keynes, MK7 6AA.

First published 1981.

Second edition 1983.

Third edition 1994.

Reprinted 1998.

Edited, designed and typeset by The Open University.

Printed in the United Kingdom by Thanet Press Limited, Margate, Kent.

ISBN 07492 51220

This text forms part of an Open University Level 2 course. If you would like a copy of *Studying with the Open University*, please write to the Course Reservations and Sales Centre, PO Box 724, The Open University, Walton Hall, Milton Keynes, MK7 6ZS, United Kingdom. If you have not enrolled on the Course and would like to buy this or other Open University material, please write to Open University Worldwide, The Berrill Building, Walton Hall, Milton Keynes, MK7 6AA.

3.2

7876C/s247b7i3.2

CONTENTS

STUDY GUIDE FOR BLOCK 7

The components of this Block are the main text and a Home Experiment. The Block, plus Case Study 4 and its associated video sequence, should occupy two study weeks. In Section 4.3.2 you are asked to examine the silicate mineral samples from the Home Kit.

You should begin your study of Block 7 at the start of Study Week 15. You should aim to reach as far as Section 5.3, including the Home Experiment, by the end of this study week. In Study Week 16, you should complete your study of Block 7 and read Case Study 4.

Your Orbit model kit will help you to visualize the structures discussed throughout your study of Block 7.

The experiments 1–12 that comprise the Home Experiment for Block 7 will take between 1 and 2 hours. Ensure that you have acquired some 20 vol. (6%) hydrogen peroxide and a gas cylinder before you start, since these items are not provided in the Home Kit. Your observations in these experiments will be tested in a TMA.

1 INTRODUCTION

In this last Block of S247, our survey of the descriptive chemistry of the typical elements will be completed. Figure 1 is a reminder of where that descriptive chemistry has been discussed in the Course. In Block 3, we dealt with Groups I and II of this mini-Periodic table, the alkali and alkaline earth metals, which lie on the left-hand side. Useful insights into their chemistry could be obtained by using an ionic model. In Block 5, we turned to the descriptive chemistry of hydrogen, the halogens and the noble gases, which lie to the right of Figure 1, in Groups VII and 0. All these elements are non-metals, and covalent compounds are much more prominent in their chemistry. Covalent bonding models therefore dominated the theoretical framework within which that chemistry was discussed. Now, in Block 7, we deal with the elements of Groups III–VI; unlike the Groups that we have discussed in Blocks 3 and 5, these contain both metals and non-metals.

Our treatment of Groups III–VI will focus especially on the elements of the second and third Periods: boron, aluminium, carbon, silicon, nitrogen, phosphorus, oxygen and sulphur. These are all elements of great social and industrial importance. The chemistry of sulphur, for example, is taken up again in Case Study 4, where its ecological impact, in the form of acid rain, is examined.

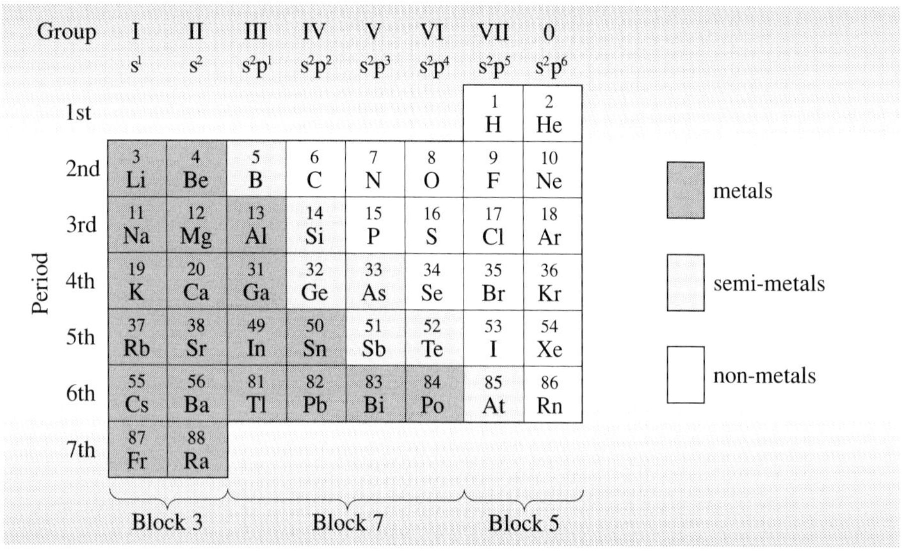

Figure 1 The plan used in S247 for the descriptive chemistry of the typical elements.

2 GENERAL OBSERVATIONS ON SECOND- AND THIRD-ROW ELEMENTS

As the introduction has shown, the continuing emphasis in Block 7 on second- and third-row elements coincides with our completion of the descriptive chemistry of the typical elements as a whole. So if we are also prepared to draw on the descriptive chemistry of Blocks 3 and 5, we shall be able to examine similarities and differences between second- and third-Period elements across the entire width of Figure 1. The relevant elements are shown in Figure 2.

We naturally expect similarities between elements in the same Group of Figure 2, such as nitrogen and phosphorus, but there are important differences as well. This subject is taken up at the end of the Block, in Section 7, but it may help you to put the intervening descriptive chemistry into perspective if we make a few general observations on it now.

						H	He
Li	Be	B	C	N	O	F	Ne
Na	Mg	Al	Si	P	S	Cl	Ar
K	Ca						

Figure 2 The elements of the second and third Periods (highlighted in grey) in the Periodic Table.

2.1 SINGLE AND MULTIPLE BONDS

One difference between second- and third-row elements is that substances containing multiple bonds are more easily made or preserved in the second row than they are in the third. Some important examples can be found among the structures of the elements themselves (Figure 3).

Group	I	II	III	IV	V	VI	VII	0
	METALLIC		MOSTLY EXTENDED STRUCTURES		COVALENT	DIATOMIC		MONATOMIC
Period 1st							H—H single bonds	He
2nd	Li	Be	B	C	N≡N	O=O multiple bonds	F—F	Ne
3rd	Na	Mg	Al	Si	P (also P_4)	S_8 rings	Cl—Cl	Ar
4th	K	Ca	Ga	Ge	As	Se_∞ chains	Br—Br	Kr
5th	Rb	Sr	In	Sn	Sb	Te_∞ chains	I—I	Xe
6th	Cs	Ba	Tl	Pb	Bi	Po	At?	Rn
7th	Fr	Ra						

Figure 3 The structure of the elemental forms of the typical elements.

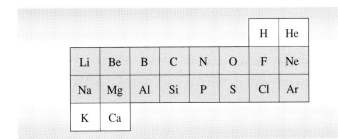

■ Why do the structures of elemental oxygen and sulphur provide such an example?

□ The second-row element oxygen consists of O_2 molecules containing a *double* bond (O=O); the commonest allotrope of sulphur, rhombic sulphur, contains S_8 rings held together by S—S *single* bonds (Figure 4). This difference was discussed in Block 5, Section 7.2.

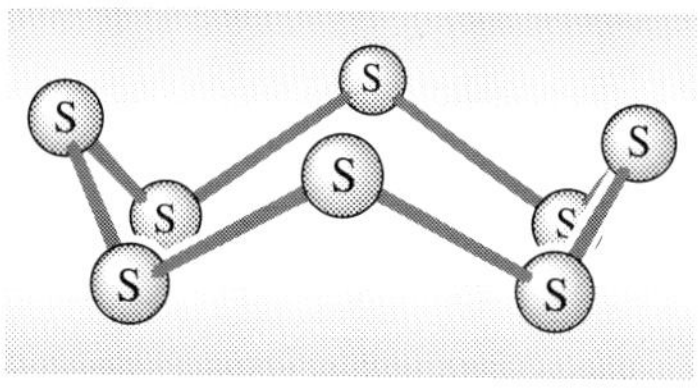

Figure 4 Rhombic sulphur, the commonest allotrope of sulphur, is a molecular solid containing S_8 molecules. Acting between these molecules to hold the solid together, there are weak van der Waals forces. The S_8 molecules are rings in which the sulphur atoms are linked by S—S single bonds.

Group IV (Figure 1) provides a further example. Graphite is the commonest form of carbon, and consists of hexagonal rings of carbon atoms, which share sides to form sheets (Figure 5; see also Block 2, Section 6).

The broken circles within the hexagons mark the delocalization of π electrons which gives rise to a bond order greater than one ($1\frac{1}{3}$). By contrast, elemental silicon occurs only in a diamond structure, with single Si—Si bonds. Nor is this difference between the second and third Periods confined to the elemental states. In Group IV, for instance, carbon dioxide consists of small discrete molecules, O=C=O, containing two carbon–oxygen double bonds. But SiO_2, the dioxide of the third-row element silicon, contains only Si—O single bonds, and this is achieved by an extended covalent structure (Figure 6), in which the coordination to the oxygens around each silicon is tetrahedral, and the coordination to the silicons around each oxygen is V-shaped.

Thus, SiO_2 is a giant molecular solid: its high boiling temperature of ≈2 800 °C reflects the strength of the Si—O single bonds that must be broken before it can volatilize as smaller units. The great differences in the volatility and physical states of CO_2 and SiO_2 are therefore a direct consequence of the preference for single bonding in the third Period, and for multiple bonding in the second. The theoretical reasons for this preference will be discussed in later Sections of the Block, when you have encountered more examples of it.

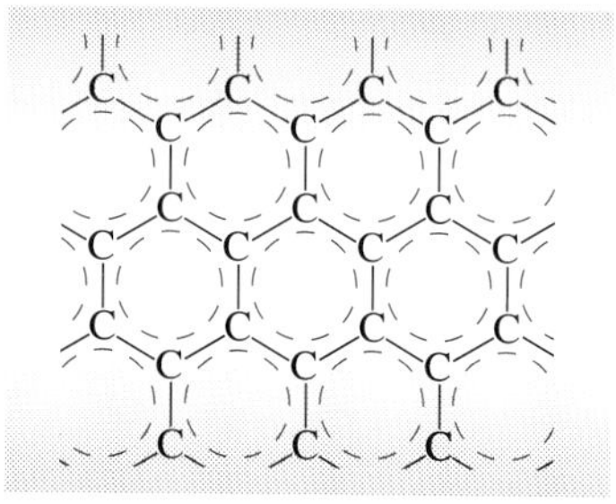

Figure 5 The structure of the sheets of carbon atoms in graphite: the broken circles mark the electron delocalization in π orbitals above and below the plane of the sheets, which leads to a carbon–carbon bond order greater than one ($1\frac{1}{3}$).

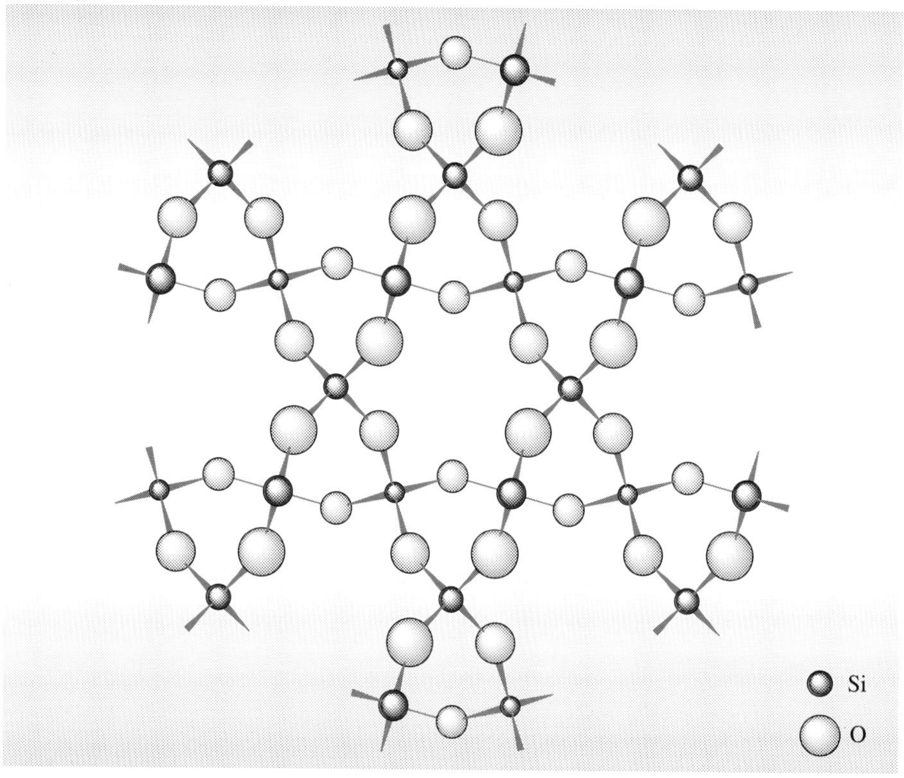

Figure 6 The structure of one form of SiO_2, β-quartz.

2.2 THIRD-ROW ELEMENTS—A CASE FOR EXPANSION OF THE OCTET?

The most elementary theories of covalent bonding are founded on two concepts, namely G. N. Lewis's two-centre electron-pair bond and the stability of the octet. Some very common compounds of the third and later rows of typical elements prove these concepts wanting.

The fluorides of nitrogen and phosphorus provide one example. The highest fluoride of nitrogen is NF_3. This is a gas at room temperature, and it contains NF_3 molecules (Structure **1**) in which N—F electron pair bonds provide both nitrogen and fluorine with octet configurations. However, phosphorus, a third-row element, forms *two* gaseous fluorides at room temperature, PF_3 and PF_5. The explanation of the bonding in PF_3 is a simple re-run of the NF_3 argument: three electron pair bonds yield octets on the fluorine and the Group V atoms. The case of PF_5 is different.

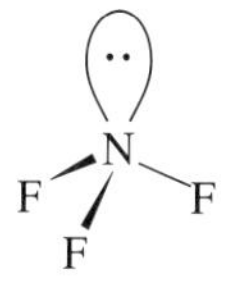

1

■ How many outer electrons does the phosphorus atom in PF_5 acquire, if it forms shared electron-pair bonds?

□ Ten; each of its five outer electrons are paired with a fluorine electron in a P—F bond.

This allocation is the one used in VSEPR theory: the repulsions between the five bond pairs of the five P—F bonds give rise to a PF_5 molecule with a trigonal bipyramidal shape (Structure **2**). Forced to choose between the octet and the two-centre electron-pair bond, we retain the latter and abandon the former.

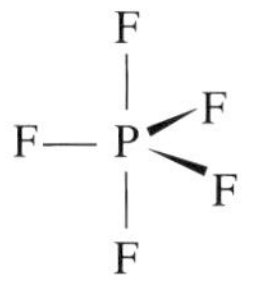

2

An important consequence of this choice emerges when we consider what phosphorus orbitals must be used to form the five electron-pair bonds. The outer electronic configuration of phosphorus is $3s^23p^3$ (Structure **3**): five unpaired electrons are needed to form the five bonds, and only three are available.

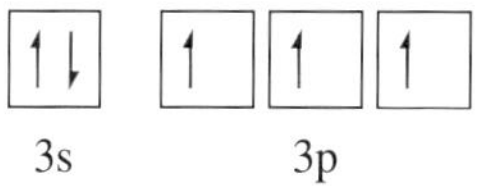

3

■ What assumption might be made to put this right?

□ The same one that we made in one of the treatments of the bonding in XeF_2 (Block 5, Section 11.4): we assume that, when bonding to fluorine, the phosphorus atom promotes one electron from its $3s^2$ pair to an empty, higher-energy orbital. The one usually chosen is one of the five 3d orbitals (Structure **4**). These normally become occupied only when we reach the next row of the Periodic Table. The energy needed to promote the electron is then recouped by forming five bonds instead of three.

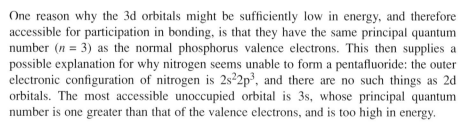

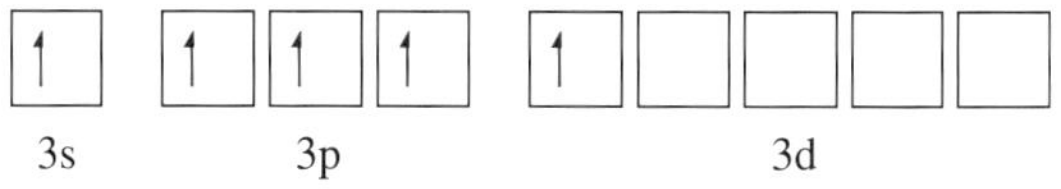

4

One reason why the 3d orbitals might be sufficiently low in energy, and therefore accessible for participation in bonding, is that they have the same principal quantum number (n = 3) as the normal phosphorus valence electrons. This then supplies a possible explanation for why nitrogen seems unable to form a pentafluoride: the outer electronic configuration of nitrogen is $2s^22p^3$, and there are no such things as 2d orbitals. The most accessible unoccupied orbital is 3s, whose principal quantum number is one greater than that of the valence electrons, and is too high in energy.

But despite this apparent plausibility, the involvement of phosphorus 3d orbitals in the bonding in PF_5 is controversial. Although these orbitals have the same n value as the $3s^23p^3$ valence electrons, they are still rather high in energy. This has prompted alternative molecular orbital treatments in which σ-bonding molecular orbitals are formed from a 2p orbital on each fluorine, and from just the 3s and three 3p orbitals on phosphorus.

■ How many molecular orbitals will be formed from such a combination?

□ Nine; there are nine atomic orbitals which must generate nine molecular orbitals.

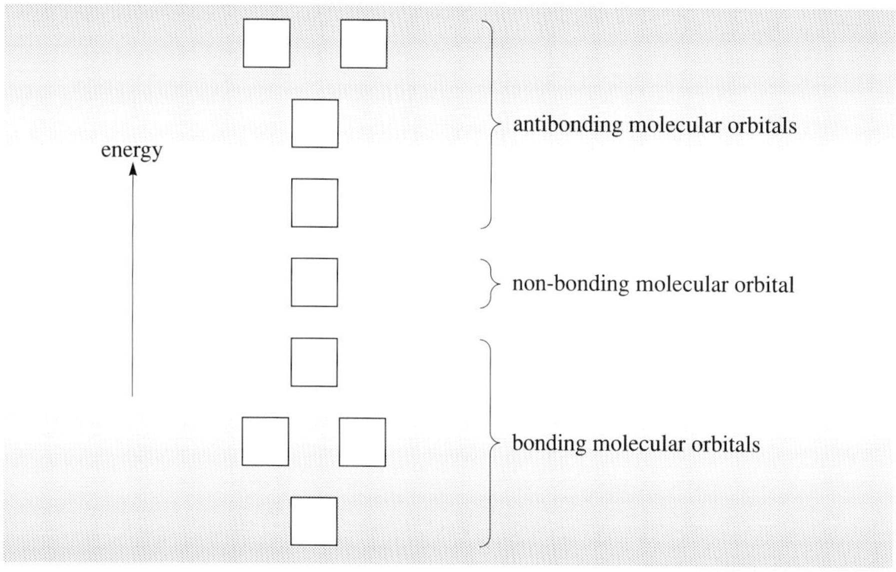

Figure 7 The sequence of σ-bonding molecular orbitals in a treatment of the bonding in PF_5, which uses a 2p orbital on each fluorine, and the 3s and 3p orbitals on phosphorus (to be filled in by students).

Four of these molecular orbitals are bonding, and there are four matching antibonding ones; this leaves one non-bonding orbital (Figure 7).

■ Pencil the appropriate number of electrons into Figure 7. What is the *average* bond order of a P—F bond according to this scheme?

□ Four-fifths: you have ten electrons (five from phosphorus and one from each fluorine 2p orbital) to assign. These fill the four bonding orbitals, and the one non-bonding orbital, giving a total bond order of four, which is distributed over five bonds.

When we choose this option then, the P—F bond order drops below one; we put forward an analogous molecular orbital treatment when we dealt with XeF_2 (Block 5, Section 11.4): the bond order per Xe—F bond was one-half, rather than one as implied by VSEPR theory.

But although the choice of bonding treatment may be a matter of opinion, there is no dispute about the experimental facts: third- and later-row elements of Groups V–VII, such as phosphorus, sulphur and chlorine, form higher fluorides and oxides than their counterparts among the second-row elements, nitrogen, oxygen and fluorine. You will see more examples of this trend as you read through Block 7.

2.3 SUMMARY OF SECTIONS 1 AND 2

1 Substances containing multiple bonds are more easily made or preserved when they contain second-row rather than third-row typical elements.

2 Third-row typical elements of Groups V–VII form higher fluorides and oxides than their counterparts in the second row.

3 The bonding in these compounds can be described as the formation of a number of conventional two-electron bonds equal to the classical valency (e.g. five in PF_5). This requires involvement of d orbitals in bonding. Alternatively, there are molecular orbital treatments that require only the s and p orbitals of the third-row element, and imply lower bond orders.

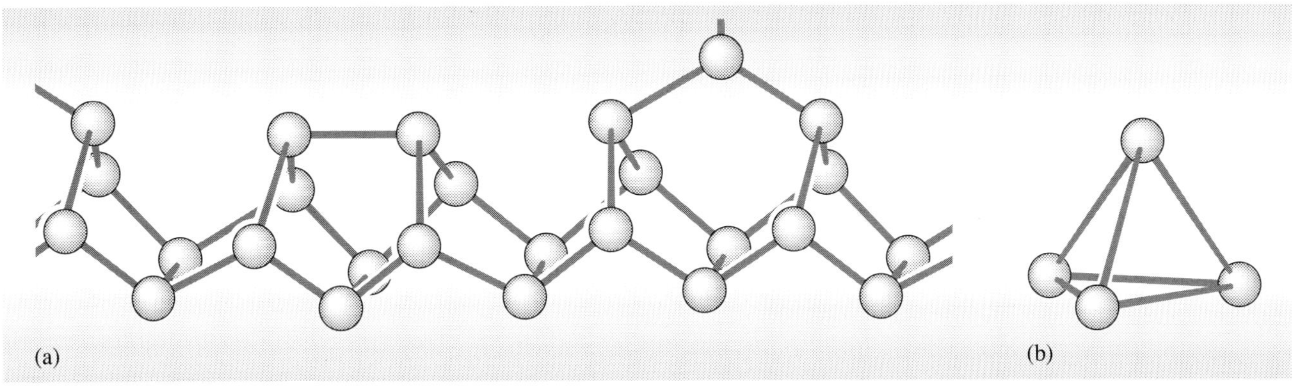

Figure 8 (a) The possible basic chain unit in the structure of red phosphorus; (b) the structure of the P_4 molecules in white phosphorus.

SAQ 1 There are two common forms of phosphorus: red and white. Red phosphorus is a solid, whose structure is not known with certainty, but the basic structural units are believed to be chains of the type shown in Figure 8a; white phosphorus is a waxy solid consisting of discrete P_4 molecules, in which the phosphorus atoms occupy the corners of a tetrahedron (Figure 8b). What covalency do the phosphorus atoms in these two structures share with each other, and with the nitrogen atoms in elemental nitrogen? How does the nitrogen case differ from the two phosphorus ones?

SAQ 2 The highest fluoride of oxygen is the gas OF_2; the highest fluoride of sulphur is the gas SF_6, which contains octahedral SF_6 molecules.

(i) If we assume that each sulphur–fluorine bond in SF_6 is a single bond, how many of the five 3d orbitals on sulphur must be involved in the bonding?

(ii) A molecular orbital treatment of the bonding in SF_6 which uses a p orbital on each fluorine atom, and just the 3s and 3p orbitals on sulphur generates four bonding orbitals, four antibonding orbitals and two non-bonding orbitals. Explain why there are ten molecular orbitals in all. What is the bond order of each sulphur–fluorine bond according to this scheme?

3 THE GROUP III ELEMENTS

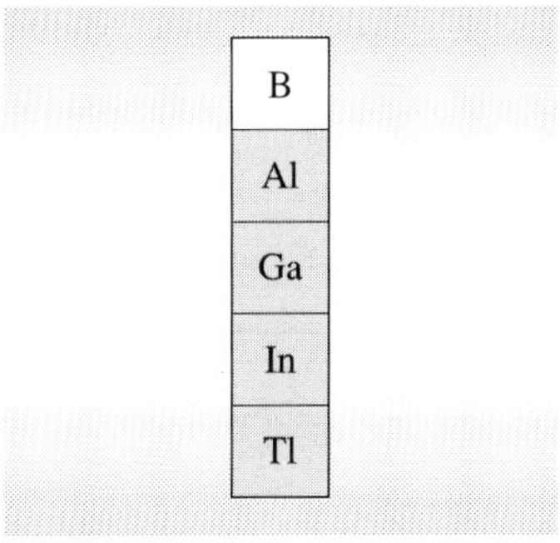

Figure 9 The Group III elements: metals are shaded grey; non-metallic boron is unshaded.

The Groups that we have tackled previously have consisted either wholly of metals (Groups I and II), or of non-metals (Groups VII and 0). Now we meet one (Figure 9) that contains both: boron is usually regarded as a non-metal (its classification in Figure 1 will be discussed in Section 3.1.3); aluminium, gallium, indium and thallium are metals.

We acknowledge this greater chemical variation within the Group by dealing separately with boron and aluminium. Then we shall take gallium, indium and thallium together, a move that can be justified not only chemically, but also by reference to electronic configurations.

■ Look at Table 1; what distinguishes the electronic configurations of gallium, indium and thallium from those of boron and aluminium?

□ There are inner d (and, in the case of thallium, f) shells of lower principal quantum number between the valence electrons and the filled shells of the preceding noble gas.

Table 1 Electronic configurations of Group III atoms

Atom	Electronic configuration
B	$[He]2s^22p^1$
Al	$[Ne]3s^23p^1$
Ga	$[Ar]3d^{10}4s^24p^1$
In	$[Kr]4d^{10}5s^25p^1$
Tl	$[Xe]4f^{14}5d^{10}6s^26p^1$

3.1 BORON: OCCURRENCE AND EXTRACTION

Boron occurs naturally as borates — substances containing oxoanions in which the oxidation number of boron is +3. There are large deposits in California and Turkey. Borax, often written $Na_2B_4O_7.10H_2O$, is a good example.* Mined borax is purified by recrystallization from water. Treatment of its hot solution with a strong mineral acid like sulphuric acid causes white, crystalline boric acid, H_3BO_3, to crystallize on cooling. However, the structure of the acid molecule (Structure **5**) shows that $B(OH)_3$ is a better formulation. When boric acid is heated, boric oxide, B_2O_3, is formed:

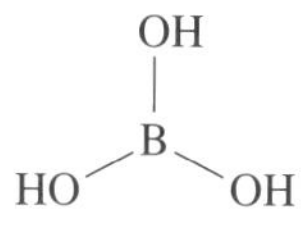

5

$$2B(OH)_3 = B_2O_3 + 3H_2O(g) \qquad (1)$$

3.1.1 THE BORON ATOM

Table 1 shows that the boron atom has three outer, valence electrons. Suppose that it uses them to form electron-pair bonds.

■ How many bonds can be formed in this way; in the process, does the atom gain a noble gas configuration?

□ It forms three single bonds, thereby increasing its number of outer electrons to six — that is, two short of the electronic configuration of neon.

This is a unique feature of boron chemistry: it is the only non-metal or semi-metal that cannot gain an octet solely by using its outer electrons to form shared electron-pair bonds. Much boron chemistry can be regarded as processes in which the atom gains a share in a greater number of electrons. Many of them can be classified as Lewis acid–Lewis base reactions.

3.1.2 LEWIS ACIDITY IN THE BORON HALIDES

The boron halides have the expected formula, BX_3. Boron trifluoride can be made by heating boric oxide, calcium fluoride and concentrated sulphuric acid:

$$B_2O_3(s) + 3CaF_2(s) + 3H_2SO_4(l) = 2BF_3(g) + 3CaSO_4(s) + 3H_2O(l) \qquad (2)$$

BCl_3 and BBr_3 can be obtained by direct combination of the elements, and BI_3 by heating BCl_3 with HI. At 25 °C, BF_3 and BCl_3 are gases, BBr_3 is a liquid and BI_3 is a solid (Table 2). All four halides are easily hydrolysed by water, boric acid being one of the products.

Table 2 Properties of the boron trihalides

	BF_3	BCl_3	BBr_3	BI_3
m.t./°C	−127	−107	−46	50
b.t./°C	−100	13	91	210
B−X distance/pm	130	175	187	210

Now from electron diffraction experiments, and according to VSEPR theory, the BX_3 molecules in these substances are trigonal planar (Structure **6**). As noted in Section 3.1.1, boron then has six outer electrons.

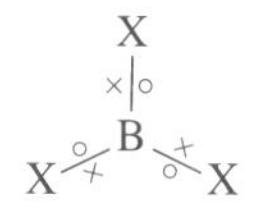

6

■ How can boron increase this number to eight (Block 5)?

* But see Section 3.1.5.

☐ The trihalide can act as a Lewis acid, acquiring a share in the lone pair of a Lewis base such as ammonia or trimethylamine; for example

$$(CH_3)_3N(g) + BX_3(g) = (CH_3)_3\overset{+}{N}{-}\overset{-}{B}X_3(s) \qquad (3)$$

In the white solid product (recall Videocassette 2, Sequence 10), the geometry around both the donating nitrogen and the acceptor boron is tetrahedral (Structure **7**); the structure of the complex is analogous to that of ethane, C_2H_6. The Lewis base may also be a halide ion; thus when BF_3 is heated with solid KF, the tetrahedral tetrafluoroborate ion (Structure **8**) is formed:

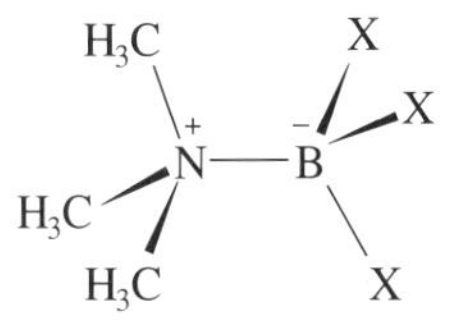

7

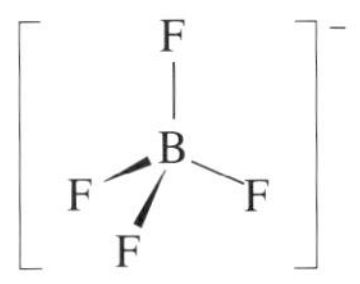

8

$$KF(s) + BF_3(g) = KBF_4(s) \qquad (4)$$

■ Using the electronegativity of the halogen as a criterion, which trihalide should form the more stable complex in Reaction 3, BF_3 or BBr_3?

☐ BF_3; the more electronegative fluorines will exercise a stronger attraction on the incoming electron pair of the Lewis base.

In fact, experiment shows the opposite: the bromide complex is the more stable; for example, $\Delta H_m^\ominus$ for Reaction 3 becomes more negative from BF_3 to BBr_3, suggesting that the stability of the $(CH_3)_3NBX_3$ complexes is in the order F < Cl < Br.

We can begin to explain the stability order by considering the bonding in the boron trihalides, which was said to involve π-bonding in Block 4 (Figure 65), and is here illustrated in Figure 10.

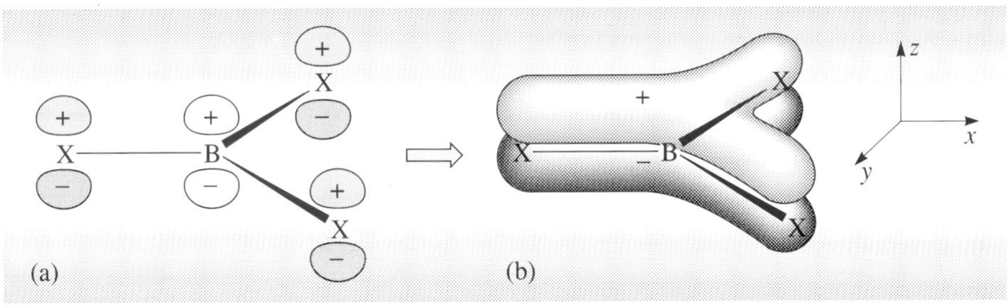

Figure 10 The bonding in the boron trihalides. The three two-electron bonds in the *xy* plane use a combination of the boron 2s, $2p_x$ and $2p_y$ orbitals. This leaves an *empty* boron $2p_z$ orbital above and below the plane as in (a). Because the molecule is planar, this overlaps with filled p_z non-bonding orbitals on the halogen atoms, resulting in partial delocalization of halogen non-bonding electrons into the boron $2p_z$ orbital. In other words, the B—X bonds are strengthened by π-bonding via a π orbital extending over all four atoms as in (b).

Evidence for the π-bonding is provided by bond lengths: in free BF_3, for example, the B—F distance is 130 pm (Table 2); in complexes like those shown in Structures **7** and **8**, where the planar geometry has been destroyed, and the possibility of π-bonding is much diminished, it is longer (≈140 pm) and so the bonds are presumed weaker. The π-bonding is believed to be most marked in BF_3, where the small fluorine atoms are closer to the central boron than in the other boron halides, which allows better overlap with the vacant boron $2p_z$ orbital. Because this π-bonding largely disappears in reactions such as Equation 3, BF_3 is more reluctant to undergo the reaction than the other trihalides, which accounts for the order of stability of the $(CH_3)_3BX_3$ complexes mentioned above: F < Cl < Br.

This Section has shown you two ways in which a trivalent boron atom can increase the number of electrons that it shares with other atoms: boron compounds can act as Lewis acids, or, if a boron atom is bound to halogen or oxygen atoms with non-bonding electrons, it can participate in π-bonding.

3.1.3 BORON AND THE BORIDES

Boron of purity 95–98% can be made by reducing B_2O_3 in a furnace with magnesium. High-purity boron can be obtained by the decomposition of the triiodide at 1 000 °C:

$$2BI_3(s) = 2B(s) + 3I_2(g) \qquad (5)$$

Boron is a hard inert black solid of very low electrical conductivity ($5 \times 10^{-5}\,S\,m^{-1}$), which melts at 2 180 °C.

Elemental boron occurs in several forms, but they all consist of interconnected B_{12} units, in which the twelve boron atoms are positioned at the corners of a regular icosahedron, a solid figure with twenty triangular faces (Figure 11).

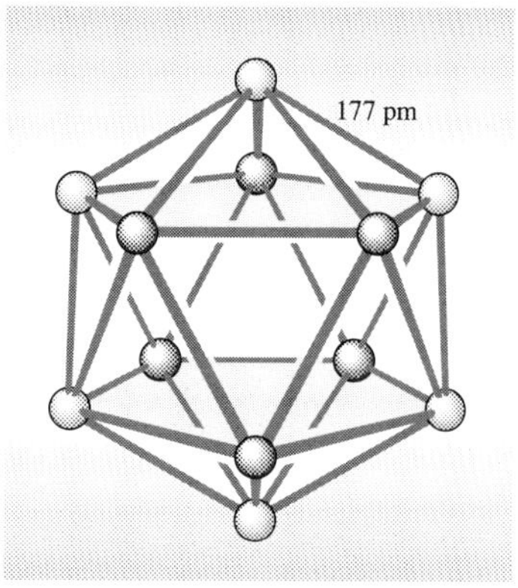

Figure 11 The different forms of elemental boron consist of interconnected B_{12} icosahedra. Each icosahedron is linked to others through the atoms shown in colour. The core of the icosahedron is a pair of parallel pentagons of boron atoms with opposite orientations; these are shaded. The icosahedron is completed by boron atoms above and below the pair. Ignoring slight irregularities in the real crystal, the unit has twenty faces, each consisting of an equilateral triangle of boron atoms of side 177 pm.

■ Could the lines between the boron atoms in Figure 11 represent conventional single bonds?

□ No; each boron is joined to *five* others, but its three outer electrons can form a maximum of three electron pair bonds.

In the simplest form of elemental boron, each B_{12} icosahedron is linked to six others by B—B bonds formed by the six coloured atoms in Figure 11. These have a length of 171 pm, nearly identical with the B—B distance in gaseous B_2Cl_4 (Structure **9**). They can therefore be regarded as orthodox single electron-pair bonds. Now each boron atom has three valence electrons, so there are thirty-six valence electrons in each icosahedron. As six of these must be used in the six B—B inter-icosahedral single bonds, that leaves thirty. These thirty must largely be dispersed within the icosahedron so as to bind the twelve boron atoms together. They may be thought of as occupying bonding molecular orbitals, which extend over the whole B_{12} unit. Alternatively, they can be regarded as an electron gas, which creates a metallic type of bonding that is localized within and around the icosahedron. Because of this 'localized delocalization', electrons cannot easily be transferred from icosahedron to icosahedron in an electric field, so boron is a non-conductor. It is because of this localized metallic character that boron has been classified as a semi-metal in Figure 1, even though its electrical conductivity is characteristic of a non-metal. Localized delocalization, or the formation of multi-centre molecular orbitals, is a further means by which boron atoms acquire a share in a larger number of electrons than that made available by the formation of localized, shared electron-pair bonds.

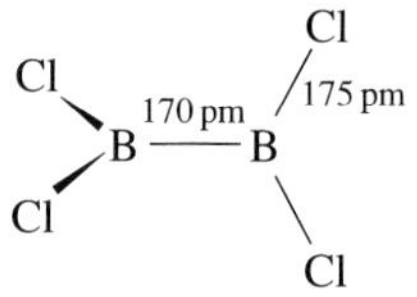

9

The localization can be broken down when boron forms metallic borides. These are often metallic conductors, and can be thought of as alloys. Titanium diboride, TiB_2, for example, is best made by heating boric and titanium oxides with carbon:

$$TiO_2(s) + B_2O_3(s) + 5C(s) = TiB_2(s) + 5CO(g) \qquad (6)$$

In this substance (Figure 12), there is very strong metallic bonding within and between layers; the material is very hard, melts close to 3 000 °C, and has a conductivity five times that of titanium metal. It is used as a protective coating for metals like tungsten in high-temperature applications.

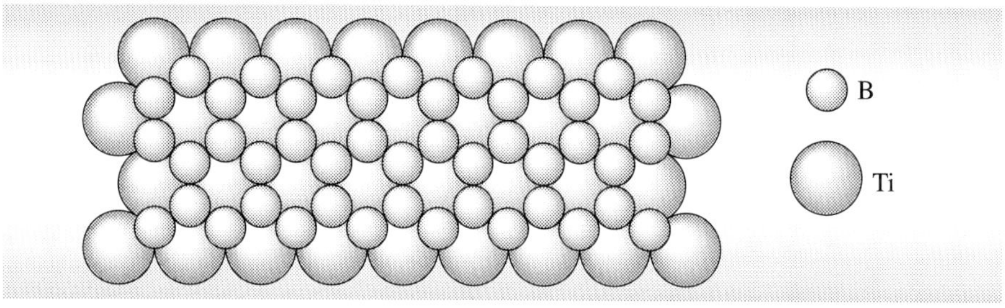

Figure 12 The structure of titanium diboride, TiB_2. Titanium has a hexagonal close-packed structure, ABABAB, etc., in which the spheres in a B-type layer sit on alternate sockets (hollows) in close-packed A-type layers. In titanium diboride, the B-type layers are replaced by layers of boron atoms, which sit on *every* socket in the titanium A-type layer. As the number of sockets in a layer is twice the number of spheres, the empirical formula is TiB_2.

3.1.4 THE BORON HYDRIDES

The extraordinary nature of boron hydrides was first revealed by the German chemist, Alfred Stock (Figure 13).

Figure 13 Virtually nothing was known about the boron hydrides in 1909, when Alfred Stock (1876–1946) began his work at what was then Breslau in Germany (now Wroclaw, Poland). As the hydrides inflame spontaneously in air, he devised the now-standard technique of vacuum line chemistry, including special grease-free valves that were necessary when handling such reactive compounds. He isolated B_2H_6, B_4H_{10}, $B_{10}H_{14}$ and, subsequently, B_5H_9, B_6H_{10} and B_5H_{11}. Liberal use was made of mercury in pumps and manometers, and during this time, he suffered headaches, vertigo and numbness, culminating, in 1923, with almost total loss of memory and hearing. Having recognized those symptoms as indications of mercury poisoning, he then spent much of the remainder of his working life on this subject, often using himself as a guinea pig, and publishing numerous valuable warnings and precautionary advice. His health deteriorated further in the 1930s, and he had political difficulties with the Nazis. By 1943, hardening of the muscles drove him to retire to Silesia, but as the Russians closed in, he and his wife became refugees; in 1946, he died in obscurity at Aken on the Elbe, after a life of much accomplishment and suffering.

■ From what you know already, what should be the molecular formula of the simplest boron hydride?

□ As there are halides, BX_3, one is entitled to expect a hydride, BH_3.

In fact, the simplest hydride to have been isolated is diborane, B_2H_6. It is made industrially by heating BF_3 with sodium hydride:

$$6NaH(s) + 2BF_3(g) = 6NaF(s) + B_2H_6(g) \qquad (7)$$

The formula is analogous to ethane, but the structure (Figure 14) is different. To interpret the bonding we assume that there are four B—H single bonds in the grey plane. This leaves one electron on each boron atom, and one electron from each of the two bridging hydrogens, a total of four in all. These four electrons are assigned to two three-centre B—H—B bonds, one being above and one being below, the grey plane. The molecular orbital treatment of these three-centre bonds resembles that used for $[F—H—F]^-$ in SAQ 33 of Block 5. The two electrons that we assign to each one occupy a bonding molecular orbital formed from a hydrogen 1s orbital and two boron orbitals (Figure 15).

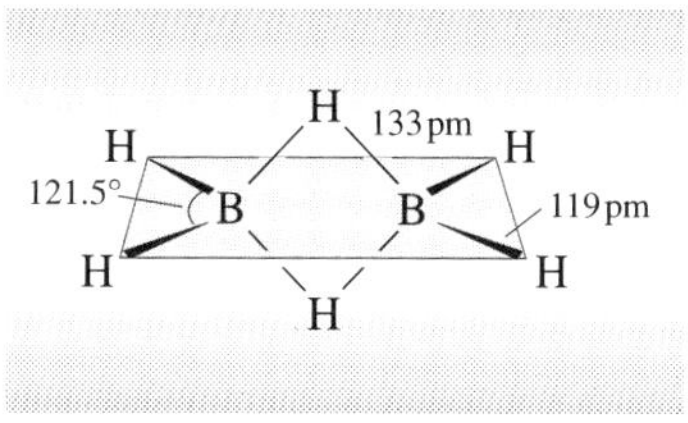

Figure 14 The structure of diborane, B_2H_6; the coordination around each boron is roughly tetrahedral.

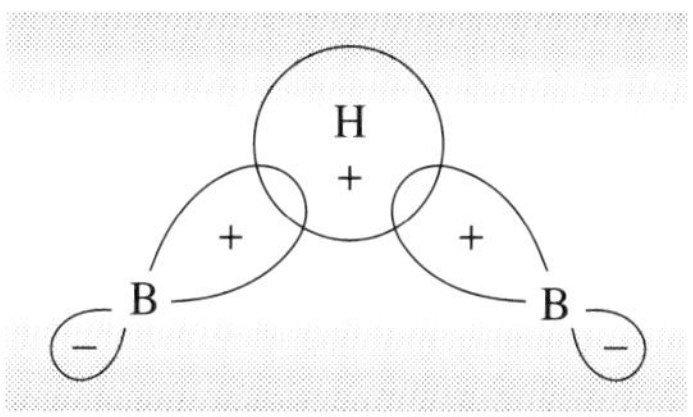

Figure 15 The atomic orbital composition of the bonding molecular orbital in one of the two BHB bent three-centre bonds that lie above and below the horizontal plane in Figure 14.

As each bonding pair is spread over two B—H distances, the B—H bond order is one-half. This is consistent with Figure 14, where the boron–hydrogen distance in the bridge is 14 pm longer than the length of the single bond in the plane.

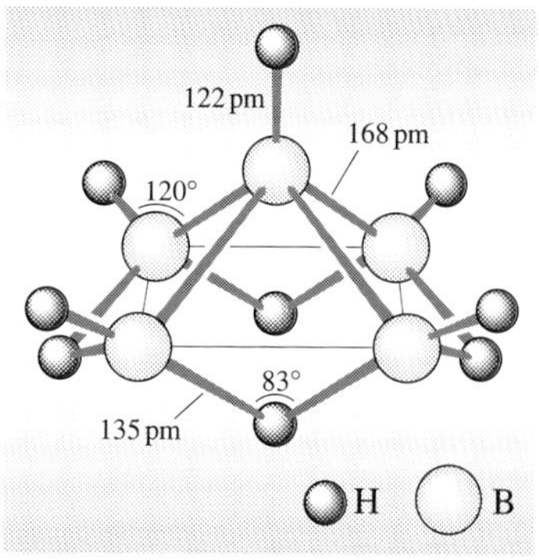

Figure 16 The structure of B_5H_9.

Diborane decomposes on heating into other, more complicated boron hydrides, and over twenty have been isolated by carefully controlling the heating conditions. For example, at 80–90 °C and 200 atmospheres, B_4H_{10} is produced; in the presence of further hydrogen, B_5H_9 is formed in a quick pass through a reaction vessel at 200–240 °C. In B_5H_9 (Figure 16), the boron atoms lie at the corners of a square pyramid, the four at the base being linked by B—H—B three-centre bonds. In addition, each boron is bound to a hydrogen atom by a conventional B—H single bond. The five borons and nine hydrogens supply twenty-four valence electrons. Ten are used in the five B—H terminal single bonds, and eight in the four B—H—B three-centre bonds. This leaves six to be delocalized over, and to hold together, the cluster of five boron atoms. Although the boron hydrides and their derivatives are many and various, they invariably feature at least two of the following: (i) B—H or B—B conventional single bonds; (ii) three-centre B—H—B bonds; (iii) multicentre bonds in a cluster of boron atoms.

3.1.5 BORON–OXYGEN COMPOUNDS

Boron–oxygen compounds account for some of the major industrial uses of boron. In boric acid, $B(OH_3)_3$, and boric oxide, B_2O_3, boron forms three coplanar, triangularly disposed bonds (Structure **10**). Crystals of boric acid are white, flaky and transparent. The $B(OH)_3$ units are linked in layers by hydrogen bonds (Figure 17); the interlayer forces are of the weak van der Waals type.

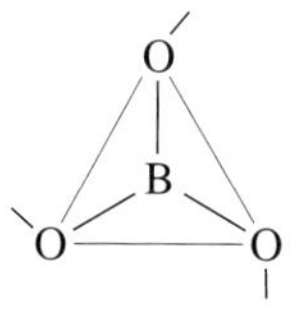

10

■ Will $B(OH)_3$ be a strong or weak acid in aqueous solution?

□ Very weak; the strength of oxoacids increases with the number of terminal oxygens in the structural formula (Block 5, Section 9.1.2), and $B(OH)_3$ has no terminal oxygens.

This is quite correct. However, the ionization of $B(OH)_3$ in water is unusual: $H^+(aq)$ is generated by abstracting OH^- from a water molecule:

$$B(OH)_3(aq) + H_2O(l) = H^+(aq) + [B(OH)_4]^-(aq) \quad (8)$$

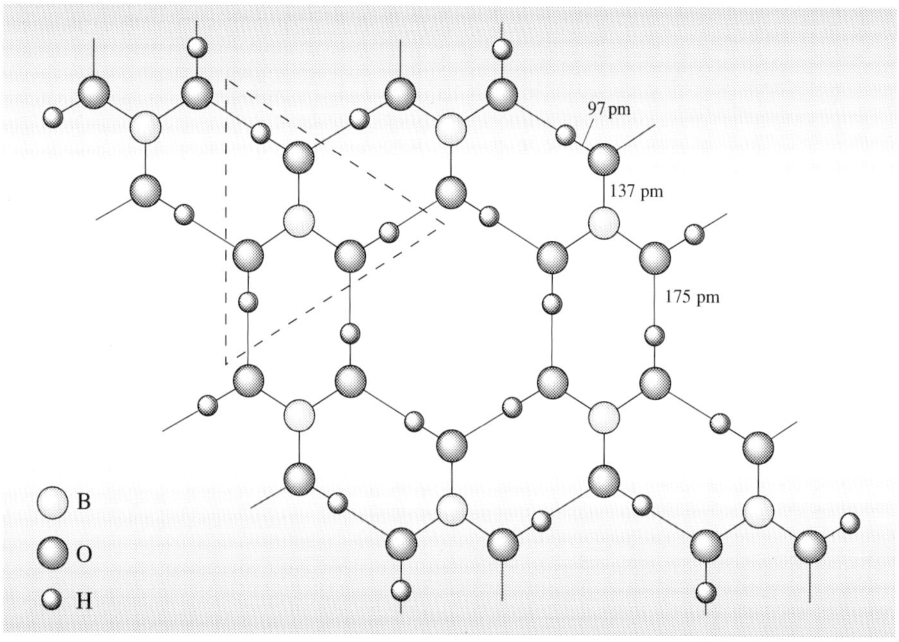

Figure 17 The structure of solid boric acid. The hydrogen bonds holding the $B(OH)_3$ units together are shown in colour. One $B(OH)_3$ unit is indicated by a broken-line triangle.

Here, boric acid acts as a Lewis acid, accepting a lone pair from the Lewis base OH^-, and forming four tetrahedrally disposed B—O bonds (Structure **11**). Thus, boron can occur in three- or fourfold coordination with oxygen atoms.

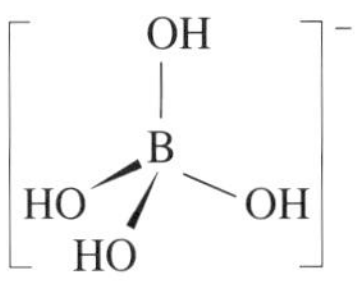

11

At normal pressures, B_2O_3 (m.t. 294 °C) consists of a three-dimensional network of BO_3 groups (Structure **10**) in which the coloured triangles share vertices. Borates can be made by opening up the B_2O_3 network using metal oxides, as indicated by a reaction such as the one shown in Figure 18. This reaction explains the use of B_2O_3 in the formation of borosilicate glasses such as Pyrex (compare Block 3, Figure 58).

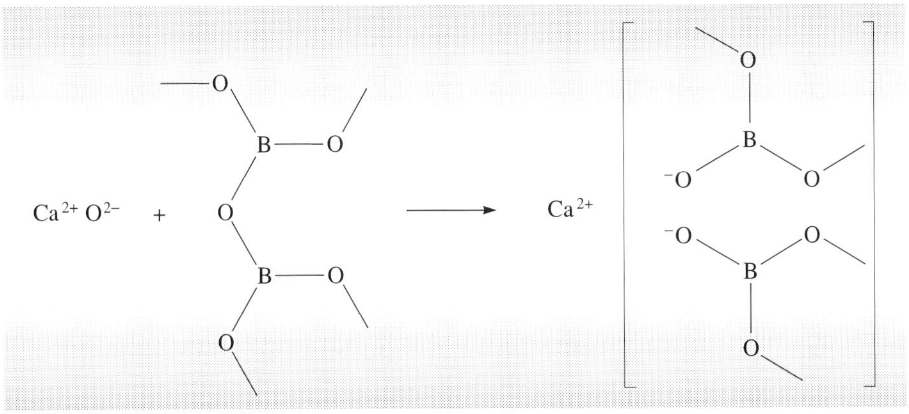

Figure 18 Metal oxides react with boric oxide to form borate glasses containing giant disordered molecular anions, or borate compounds. The reaction resembles that with silica, or with the hypothetical sheet-like oxide, G_2O_3, in Block 3, Figure 58. An O^{2-} ion is transferred to the B_2O_3 network, opening up a B—O—B link, and replacing it with two B—O^- bonds.

When the proportions of the oxides are right, crystalline metal borates are formed. If the network is fully opened up, discrete BO_3^- ions are formed (Figure 19a).

Progressively less opening is marked by pairs, rings and chains of BO_3 triangles, joined at vertices in all cases (Figure 19b, c and d). All these compounds contain just three-coordinate boron, but in some cases, four-coordinate atoms may be present as well. Thus, in the mineral borax, written $Na_2B_4O_7.10H_2O$ in Section 3.1, the anion is $[B_4O_5(OH)_4]^{2-}$ (Figure 20), in which there are two three-coordinate and two four-coordinate boron atoms.

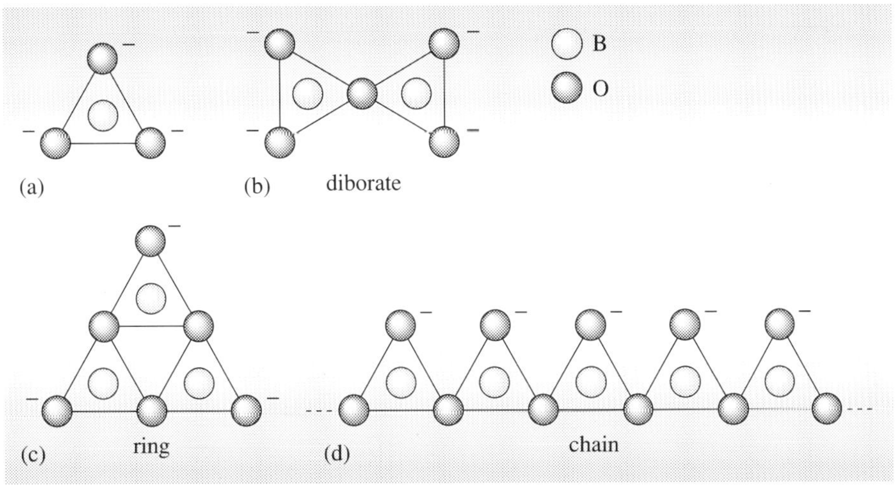

Figure 19 Borate anions that can be made by opening up the B_2O_3 network with metal oxides. These examples contain only three-coordinate boron. BO_3 triangles share zero, one or two vertices with —O^- sites at any vertex that is unshared: (a) discrete BO_3^{3-} ions (no shared vertices); (b) diborate, $B_2O_5^{4-}$, ions (each BO_3 triangle shares one vertex); (c) $B_3O_6^{3-}$ rings (each BO_3 triangle shares two vertices); (d) as (c) but the triangles are arranged in a chain.

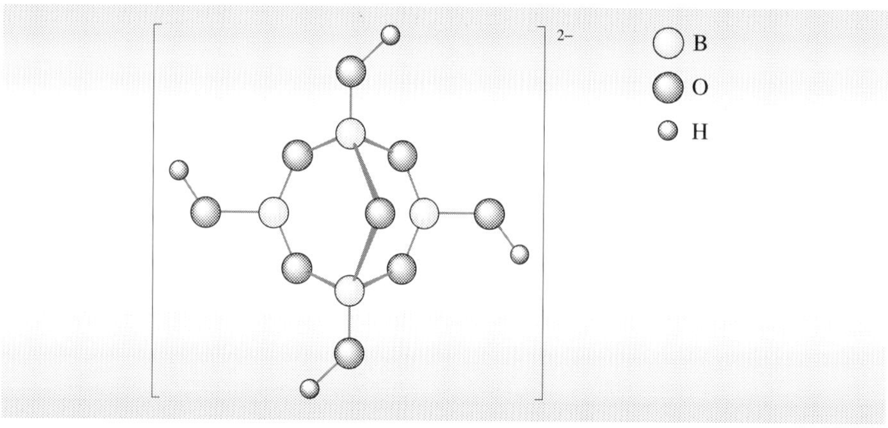

Figure 20 The ion $[B_4O_5(OH)_4]^{2-}$, which is found in the mineral borax.

■ What would be a more correct formula for the compound?

□ $Na_2[B_4O_5(OH)_4].8H_2O$; two of the ten water molecules in the usual formula must be incorporated into the borate anion.

3.1.6 SUMMARY OF SECTION 3.1

1 Boron occurs naturally as borax, $Na_2[B_4O_5(OH)_4].8H_2O$, in which boron is both trigonally and tetrahedrally coordinated by oxygen. Sulphuric acid precipitates boric acid from an aqueous solution of borax, and boric acid can be thermally decomposed to boric oxide, B_2O_3.

2 Boron falls short of an octet when it uses its valence electrons to form electron-pair bonds. In the planar halides, BX_3, the number of shared electrons is enlarged by π-bonding between the empty boron $2p_z$ orbital, and filled non-bonding orbitals on the halogens. Alternatively, boron can achieve an octet of electrons when the halides act as Lewis acids; this reaction is least favourable for BF_3, where the weakening of the π-bonding in the halide is usually the largest.

3 Elemental boron is a black, high-melting solid made by magnesium reduction of B_2O_3, or by heating BI_3. It consists of interconnected B_{12} icosahedra. There is localized metallic-type bonding (multicentre molecular orbitals) within the icosahedra, but not between them, so boron itself is a non-conductor. Borides (e.g. TiB_2), however, are often metallic.

4 In B_2H_6, two BH_2 groups are linked by bridging hydrogens in two BHB, three-centre, two-electron bonds. Other boron hydrides such as B_5H_9 contain BHB bonds, but often also involve multicentre bonds in clusters of boron atoms.

5 Boric acid is a weak acid, which generates protons by abstracting OH^- from water. B_2O_3 consists of triangular BO_3 groups linked through their vertices. Metal oxides open up the network, replacing B—O—B linkages by two $B—O^-$ bonds. This can give rise to glasses, or to borates, in which BO_3 groups are joined at only zero, one or two vertices. Some borates also contain four-coordinate boron.

SAQ 3 The halides BX_3 can be isolated as chemical compounds; the hydride BH_3 cannot. What stabilizing factor, usually assumed to be present in the boron halides, would be missing in BH_3?

SAQ 4 The structure of the hydride B_4H_{10} is shown in Figure 21. The B(1)–B(3) distance is 171 pm; the B(2)–B(4) distance is 280 pm. Identify the B—B single bonds, B—H single bonds and BHB three-centre bonds, and then find out if your selection uses up all of the valence electrons in the compound.

SAQ 5 (a) Represent the structure of B_2O_3 by using triangles of the type used in Figure 19. Explain why, even though the triangles consist of one boron atom surrounded by three oxygens, the formula is B_2O_3.

(b) The borate whose anion has the chain structure shown in Figure 19d is made by heating CaO/B_2O_3 mixtures in the right proportions. What is its empirical formula?

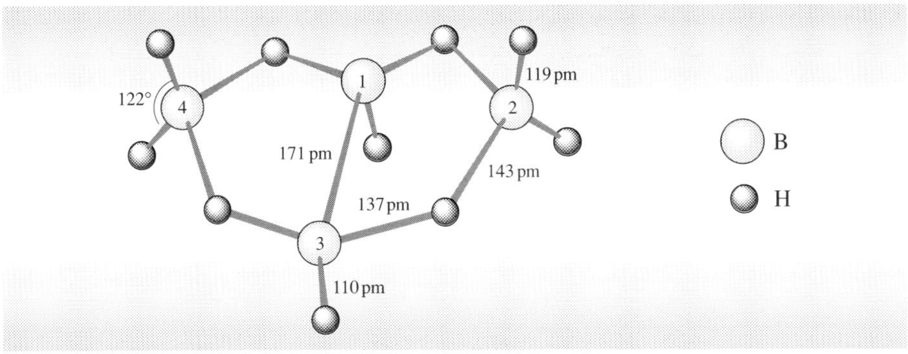

Figure 21 The structure of the hydride B_4H_{10}.

3.2 ALUMINIUM

Aluminium is the most abundant metallic element in the Earth's crust. It is extracted from bauxite, an impure hydrated oxide, which may be represented as $Al_2O_3.3H_2O$. In Block 1, Section 3.3, the purification of bauxite, and the subsequent electrolytic extraction of the metal were described. Many of the uses of aluminium — in canning, aircraft construction, and overhead transmission lines — depend on its low density and resistance to corrosion in air and water. In Block 1, Section 13, you saw that this resistance is a kinetic effect, caused by a coherent oxide film, which protects the underlying metal. Some of the important aluminium chemistry that we discuss is summarized in Figure 22.

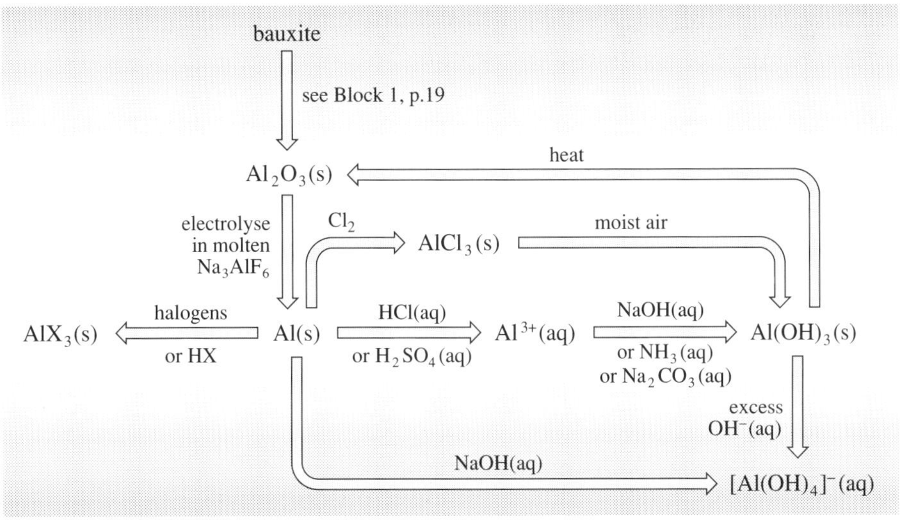

Figure 22 Some basic aluminium chemistry.

3.2.1 AQUEOUS CHEMISTRY

Like most metals, aluminium, in contrast to the non-metal boron, forms an aqueous cation. Hydrates of aluminium sulphate, $Al_2(SO_4)_3$, for example, can be made by dissolving bauxite in sulphuric acid and evaporating the solution. When such sulphates are dissolved in water, Al^{3+}(aq) is formed:

$$Al_2(SO_4)_3(s) = 2Al^{3+}(aq) + 3SO_4^{2-}(aq) \qquad (9)$$

Careful addition of aqueous sodium hydroxide to this solution will first precipitate insoluble aluminium hydroxide:

$$Al^{3+}(aq) + 3OH^-(aq) = Al(OH)_3(s) \qquad (10)$$

The oxide is produced by filtering and heating the hydroxide:

$$2Al(OH)_3(s) = Al_2O_3(s) + 3H_2O(g) \qquad (11)$$

Both oxide and hydroxide are unusual in being amphoteric. Thus, they will dissolve in and neutralize acids:

$$Al(OH)_3(s) + 3H^+(aq) = Al^{3+}(aq) + 3H_2O(l) \qquad (12)$$

■ What else will an amphoteric oxide/hydroxide do?

□ It will dissolve in and neutralize alkalis.

In this case, if *excess* sodium hydroxide is added to the precipitate that is initially formed in Reaction 10, the precipitate dissolves to form the tetrahydroxyaluminate ion:

$$Al(OH)_3(s) + OH^-(aq) = [Al(OH)_4]^-(aq) \qquad (13)$$

As Figure 22 implies, the bases ammonia and sodium carbonate are not strong enough to bring about this dissolution. The resemblance of $Al(OH)_3$ to $Be(OH)_2$, which is also amphoteric, is very close (Block 3, Section 13.2).

Because the protective oxide film on the metal is soluble in both acids and alkalis, aluminium, again like beryllium, dissolves in both hydrochloric acid and sodium hydroxide, liberating hydrogen and giving clear, colourless solutions:

$$2Al(s) + 6H^+(aq) = 2Al^{3+}(aq) + 3H_2(g) \qquad (14)$$

$$2Al(s) + 6H_2O(l) + 2OH^-(aq) = 2[Al(OH)_4]^-(aq) + 3H_2(g) \qquad (15)$$

Note, however, that in acids containing a strongly oxidizing anion, like nitric acid, HNO_3, or chromic acid, H_2CrO_4, the oxide film is maintained and no reaction occurs; the aluminium is said to be rendered passive.

3.2.2 ALUMINIUM SULPHATE AND WATER TREATMENT

In Block 3, you saw that the Group I and Group II metals formed carbonates, those of Group II being insoluble in water. Aluminium carbonate, $Al_2(CO_3)_3$, however, cannot be prepared; if aluminium sulphate is added to a solution containing carbonate or hydrogen carbonate ions, $Al(OH)_3$ is precipitated and carbon dioxide is evolved:

$$2Al^{3+}(aq) + 3CO_3^{2-}(aq) + 3H_2O(l) = 2Al(OH)_3(s) + 3CO_2(g) \qquad (16)$$

$$2Al^{3+}(aq) + 3HCO_3^-(aq) = Al(OH)_3(s) + 3CO_2(g) \qquad (17)$$

Large quantities of aluminium sulphate are used in the water industry to clear water of fine suspensions, such as clay particles, which are otherwise difficult to filter off. The tiny particles usually carry surface negative charges, which repel each other and prevent coagulation. The positively charged aluminium ions get between the negative particles, counteracting the repulsion and encouraging aggregation. Then, when Reaction 17 occurs because of the HCO_3^- ions usually present in natural waters (Block 3, Section 12.3), the particles are carried down with the precipitate of aluminium hydroxide that is formed.

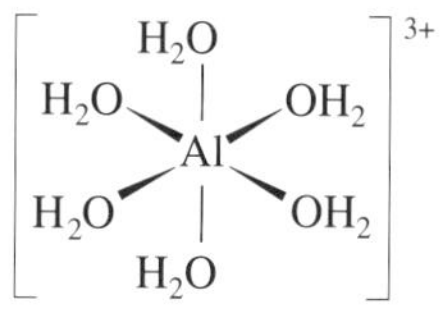

12

Reactions 16 and 17 are of interest because acids liberate CO_2 from carbonate and hydrogen carbonate solutions (Block 3, Section 7.3.1), so here, $Al^{3+}(aq)$ plays the part of an acid. This becomes more apparent when you consider that $Al^{3+}(aq)$, like $Mg^{2+}(aq)$ (Block 3, SAQ 21), is shorthand for an octahedral aquo complex (Structure **12**). Then, Equation 17 becomes

$$[Al(H_2O)_6]^{3+}(aq) + 3HCO_3^-(aq) = Al(OH)_3(s) + 3CO_2(g) + 6H_2O(l) \qquad (18)$$

■ Why can $[Al(H_2O)_6]^{3+}$ be said to act as a Brønsted–Lowry acid in this equation?

□ Three of the six water molecules that were attached to the aluminium have been lost, but the other three have acted as *proton donors*, leaving aluminium associated with hydroxide ions rather than with water molecules.

A more obvious sign of the acid character of $[Al(H_2O)_6]^{3+}$ is the fact that aqueous solutions of aluminium sulphate, unlike those of, say, sodium sulphate, are acid:

$$[Al(H_2O)_6]^{3+}(aq) + H_2O(l) = [Al(H_2O)_5(OH)]^{2+}(aq) + H_3O^+(aq) \qquad (19)$$

Again, a water molecule coordinated to Al^{3+} is transformed into an OH^- ligand; at the same time, hydrated protons are generated.

Reaction 19 was primarily responsible for the disaster at Camelford in Cornwall in July 1988. A relief driver arrived at the Lowermoor water treatment works with 20 tonnes of aluminium sulphate solution, and the key to the storage tank where it was to be deposited. Unfortunately, the key also fitted another tank where the water was held prior to discharge into the public supply mains, and it was into this tank that the driver discharged his load. Around 9 p.m. on the same evening, South West Water began receiving complaints: milk curdled when it was added to tea or coffee, because the aluminium sulphate coagulated milk in the way that it coagulates suspensions in natural waters. By the time the problem had been diagnosed, 30 000 fish had been killed in the rivers Camel and Allen, and the aluminium concentrations in Camelford drinking water briefly reached 200 times the EC limit (Figure 23).

The major problem, however, seems to have arisen not from the aluminium, but from the acidity produced in Equation 19. This dissolved copper hydroxide and carbonate that had been deposited in water pipes, thereby raising copper concentrations to seven times the EC limit. It was this copper pollution that was later blamed for mouth ulcers, sore throats, skin rashes and the occasional case of green hair (Figure 24).

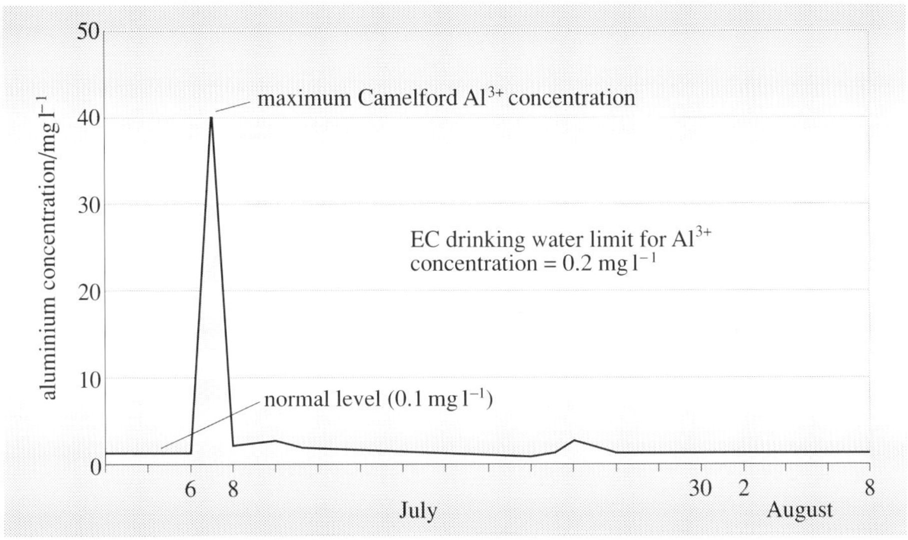

Figure 23 The variation in the aluminium concentration of Camelford drinking water in July and August 1988.

Where green hair is all the rage

By Paul Stokes

BLONDES were furious yesterday about acid tap water which turned their hair a stunning shade of green.

Dozens of them in North Cornwall have complained to the water board. Other people are up in arms about blue water which they say has caused mouth ulcers and sore throats.

Mrs Anne Reynolds, of Tintagel, said: "I had my hair highlighted a week before and was drying it with a towel when my mother rang to warn me not to wash my hair because she had heard about this problem.

"Unfortunately she was too late. My highlights had turned dark green. I was absolutely horrified. It took me five goes to wash it out."

Mrs Reynolds said: "One woman whose hair was totally blonde is now completely green."

Tracey Cobbledick, a teenager from Valley Truckle, near Camelford, found her shoulder-length hair had changed colour after she took a bath.

A middle-aged woman from Cambridge staying at a nearby caravan park cut short her holiday after a similar incident.

South West Water blamed the green hair on a breakdown at the Lowermoor treatment works, which left the water in a highly acidic state.

A spokesman assured consumers that the offending water was being flushed out and that the problem would soon be solved.

But in the meantime, 74 customers have signed a petition organised by a shopkeeper, Mrs Barbara Luke, demanding £10 off their bills.

Mrs Hilary Hughes, 42, of Treveighan, said: "I was lying in the bath when the soap reacted with the water and turned it blue instantly.

"I jumped out straight away as it gave me quite a shock. I have not had a bath since and we have all been drinking spring water.

"I think it's scandalous that the water authority did not warn people. They put warnings out on the local radio, but not everyone listens to that."

The water board spokesman did admit that the acid had corroded copper pipes, with unfortunate consequences.

"The water supply in this area was undrinkable," he said, "but not a health risk. It just tasted awful, although people with sensitive skins might have got a burning sensation.

"We have heard of bath water turning blue. This was not dangerous."

He added: "We had about 20 complaints yesterday, but it is virtually cleared up now."

Figure 24 National newspaper report following the pollution of Camelford water.

3.2.3 ALUMINIUM HALIDES

Aluminium forms solid trihalides when it is heated with the halogens or hydrogen halides. Unlike the boron trihalides, these substances do not contain discrete MX_3 molecules at room temperature. Thus, AlF_3 melts at 1 290 °C and has a three-dimensional structure, characteristic of an ionic compound, in which each aluminium is surrounded by six fluorines, and each fluorine is nearly linearly coordinated to two aluminiums. $AlCl_3$, however, has a layer structure (Figure 25), which resembles that of $MgCl_2$ (Block 3, Section 13.3).

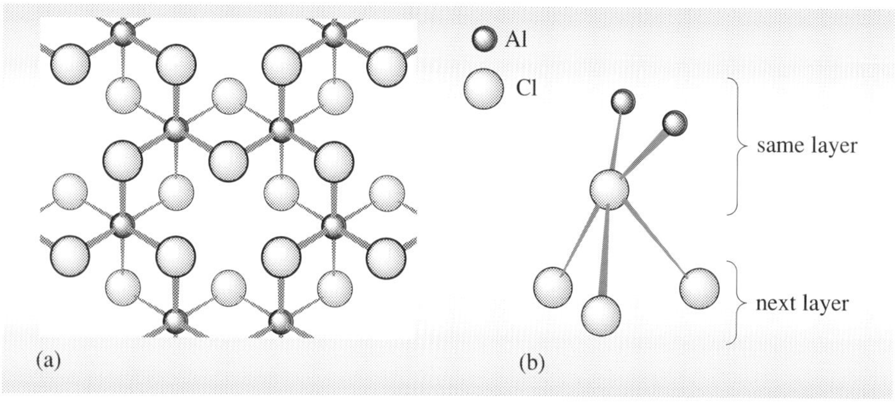

Figure 25 (a) One of the layers of the $AlCl_3$ structure viewed from above; the layer has three decks: the top deck consists of chlorines with heavy-bordered grey circles, the middle deck of octahedrally coordinated, coloured aluminiums, and the bottom deck of chlorines with light-bordered grey circles. (b) The environment of each chlorine atom: there are two aluminiums on one side, but on the other side there are three chlorines much further away in an adjacent layer.

Here the halogen environment is very different from what would be expected in a structure composed of ions. The trichloride melts under slight pressure at only 194 °C, and both the liquid and vapour contain discrete Al_2Cl_6 molecules with a structure similar to B_2H_6 (Figure 26).

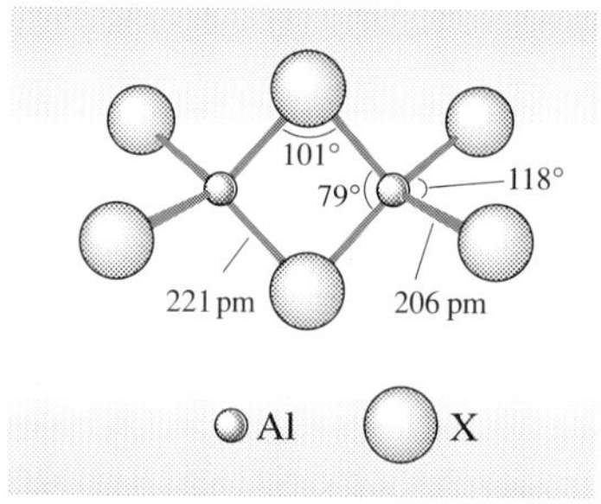

Figure 26 Structure of the Al_2X_6 molecules found in liquid and gaseous $AlCl_3$, and in solid $AlBr_3$ and AlI_3. The bond length and angle data refer to the chloride.

■ How does the percentage difference in the lengths of the terminal and bridge bonds compare with that in B_2H_6?

□ It is much smaller; in Al_2Cl_6 the bridge bond is longer by about 7%; in B_2H_6 the difference is about 12%.

This difference means that a three-centre bond treatment in which the bridge bonds have order one-half is less appropriate for Al_2Cl_6; the molecule is better represented by allowing each bridging chlorine to form one shared electron-pair bond, and one dative bond as in Structure **13**; all bonds then have order one. A bridging chlorine can do this because it can use more than one of its orbitals to form the two bonds in the bridge; hydrogen, by contrast, has only one orbital available — the 1s.

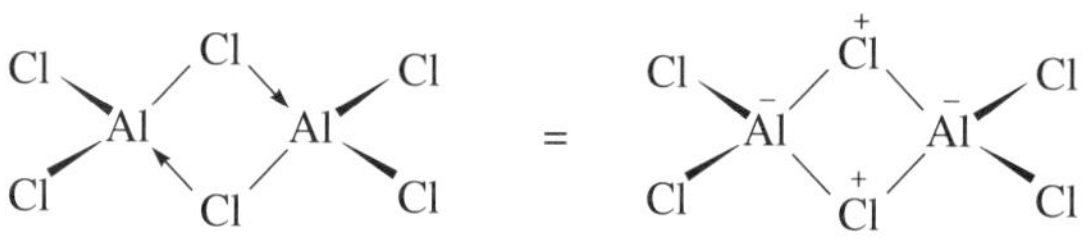

13

Solid $AlBr_3$ and AlI_3 both consist of Al_2X_6 molecules of the type shown in Figure 26. Thus, from AlF_3 to AlI_3, we see signs of increasing covalent character in the transition from a three-dimensional ionic structure in AlF_3, to a layer structure in $AlCl_3$, and then to molecular structures in $AlBr_3$ and AlI_3.

3.2.4 TWO OBSERVATIONS ABOUT ALUMINIUM CHEMISTRY

All the compounds and complexes that we have considered so far show that aluminium chemistry is dominated by just one oxidation number, +3, the value equal to the Group number. No compounds in other oxidation numbers have been characterized at room temperature. However if the metal and trichloride are heated in a sealed container to 1 000 °C, a gaseous monochloride, AlCl, is formed:

$$2Al(l) + AlCl_3(g) = 3AlCl(g) \qquad (20)$$

On cooling, the compound disproportionates, regenerating the metal and trichloride.

A second general feature of aluminium chemistry is the tendency of the element to form the oxide or hydroxide compounds. This is especially clear when comparisons are made with, say, sodium. The tendency is made evident in the insolubility of $Al(OH)_3$ in water and its solubility in alkali, in the way in which the non-existent carbonate decomposes to Al_2O_3, and in the acidity of the aqueous ion (Equation 19). Again, unlike NaCl, $AlCl_3$ fumes in moist air because hydrogen chloride gas is formed along with aluminium hydroxide:

$$AlCl_3(s) + 3H_2O(g) = 3Al(OH)_3(s) + 3HCl(g) \qquad (21)$$

Finally, the strong tendency of the metal to form the oxide provides the protective film and corrosion resistance without which the metal would be much less useful.

3.3 GALLIUM, INDIUM AND THALLIUM

In Groups I and II, sodium and magnesium are less readily oxidized than the metals beneath them. One sign of this is the less negative values of $E^{\ominus}(Na^+|Na)$ and $E^{\ominus}(Mg^{2+}|Mg)$ (Block 3, Tables 3 and 10) compared to the elements below them. In Group III, this trend is reversed: Table 3, shows that, thermodynamically speaking, aluminium is more readily oxidized to oxidation number +3 than either gallium, indium or thallium.

Table 3 Values of $E^{\ominus}(M^{3+}|M)$ for the Group III metals

	$E^{\ominus}(M^{3+}|M)/V$
Al	−1.68
Ga	−0.53
In	−0.34
Tl	+0.72

A possible reason for the reversal emerges from a comparison of the first ionization energies of the Group II and Group III elements (Figure 27).

From boron to aluminium, there is the usual drop from the second row to the third row of the Group, but thereafter, the values remain unexpectedly high, most notably at thallium whose ionization energy exceeds that of aluminium. In Block 3, Sections 3–3.2, you saw that there is a steep drop in ionization energy when a new Period begins, followed by an *overall* increase across a Period as the nuclear charge builds up.

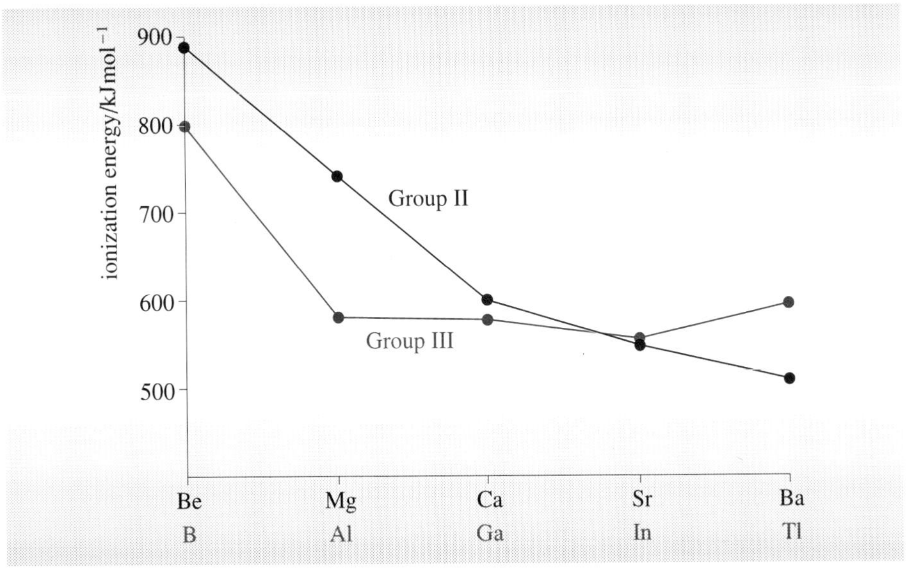

Figure 27 The first ionization energies of the Group II and Group III elements. There is a marked decrease between magnesium and barium, which is not matched between aluminium and thallium.

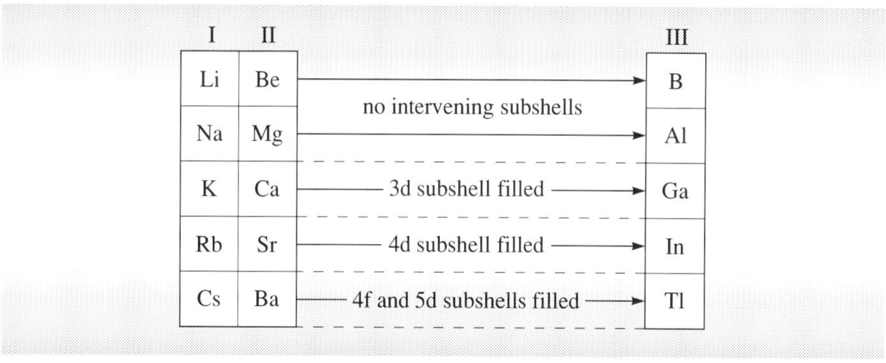

Figure 28 In the lithium and sodium Periods, no intervening subshells are filled between Groups II and III; in the potassium and rubidium Periods, however, the 3d and 4d subshells must be filled between these Groups, and in the caesium Period, both the 4f and 5d are filled.

■ Now look at Figure 28. Why might the ionization energies of gallium, indium and, especially, thallium be unexpectedly raised relative to those of aluminium?

□ Aluminium follows immediately after a Group II element, but prior to gallium and indium, d shells must be filled first. This leads to a more prolonged build-up of nuclear charge, so the ionization energies of gallium and indium are raised. The effect is magnified at thallium, where prior filling of 5d *and* 4f shells occurs.

Ionization energies strongly influence the Mulliken electronegativity (Block 3, Section 3.4), so the implication is that gallium, indium and thallium will also be more electronegative than they would be if the inner d and/or f shells were absent. The higher ionization energies also contribute to the greater resistance to oxidation revealed in Table 3.

Such effects, however, do not obliterate the strong resemblance of gallium, indium and thallium to aluminium that their presence in the same Group implies. All three elements are metals, which react with fluorine or chlorine to form trihalides, all of which are solids at room temperature. The metals also dissolve in dilute acids, evolving hydrogen and forming aqueous ions. With gallium and indium, these ions are Ga^{3+}(aq) and In^{3+}(aq); we return to thallium in a moment. Addition of alkali precipitates colourless $Ga(OH)_3$, which is amphoteric, and $In(OH)_3$, which is not.

■ What will be the effect of adding excess NaOH to the two suspensions?

□ $In(OH)_3$ will be unaffected; $Ga(OH)_3$ dissolves in the following reaction:

$$Ga(OH)_3(s) + OH^-(aq) = [Ga(OH)_4]^-(aq) \quad (22)$$

In descending Group III from B_2O_3 to In_2O_3, there is therefore a decrease in acidity, and an increase in the basic character of the oxides/hydroxides: B_2O_3 and $B(OH)_3$ are acidic, Al_2O_3 and Ga_2O_3 are amphoteric, and In_2O_3 is basic. Tl_2O_3 confirms this trend, since it too is basic.

Finally, a few comments on thallium. Look at Table 3, remembering that $E^\ominus(H^+|\frac{1}{2}H_2) = 0$ V.

■ When thallium metal dissolves in dilute acids, with evolution of hydrogen, can the product be Tl^{3+}(aq)?

□ No; $E^\ominus(Tl^{3+}|Tl) > E^\ominus(\frac{1}{2}H_2|H^+)$; thallium going to Tl^{3+}(aq) cannot reduce H^+(aq) to hydrogen gas.

In fact, the product is Tl^+(aq); thallium, more than any other Group III metal, has a prominent +1 oxidation number. The build-up of nuclear charge in the preceding 4f and 5d block elements leaves thallium's ionization energies higher than they would otherwise be. The higher oxidation number is therefore harder to attain, and the state most stable to oxidation or reduction is +1.

■ Write down the electronic configuration of Tl^+.

□ [Xe] $4f^{14}5d^{10}6s^2$; the outer 6p electron has been lost, leaving two outer electrons in a full 6s shell.

(a)

(b)

Figure 29 Two notorious thallium poisoners: (a) The amiable-looking Caroline Grills despatched four relatives and family friends, and blinded another, in Sydney, Australia during the late 1940s. During the subsequent life-sentence, her fellow inmates christened her 'Aunt Thally'. (b) Graham Young, whose face fits the part rather better, had already done time for poisoning in Broadmoor when, in 1971, he began doctoring his workmates' tea. Two men who each received a total of about 1 g of thallium(I) acetate, subsequently died, and the police found that Young had kept a careful record of the doses and symptoms of these and other victims. He was returned to Broadmoor, where he died in 1990.

The emergence at the bottom of Groups III–V of a stable lower oxidation number, two less than the Group number is sometimes called the **inert pair effect**, because the outer electronic configuration of the ion is a filled s^2 subshell, which is presumed hard to remove during oxidation. The effect increases down the Group: AlCl, AlBr and AlI do not exist at room temperature, but the corresponding compounds of gallium and indium can be made by cooling a heated mixture of the metals and their trihalides:

$$2M + MX_3 = 3MX \qquad (23)$$

They all, however, decompose in water, evolving hydrogen, or disproportionating to the metal and M^{3+}(aq). Only in the case of thallium does a long-lived M^+(aq) ion exist. The ionic radius of Tl^+ (160 pm) resembles that of K^+, and, like potassium, thallium forms, unusually, a soluble, alkaline carbonate and hydroxide, Tl_2CO_3 and TlOH. When ingested, thallium seems to follow potassium in its metabolism, and it probably interferes with vital roles played by potassium in the nervous system. Certainly thallium, and especially thallium(I), is extremely poisonous, as some well-publicized murder cases testify (Figure 29).

3.4 SUMMARY OF SECTIONS 3.2 AND 3.3

1 Al_2O_3 and $Al(OH)_3$ are amphoteric. Thus, the protective film on the metal can be dissolved by both acids and strong alkalis, so the metal dissolves in either aqueous NaOH or HCl, with evolution of hydrogen.

2 The strong tendency of aluminium to become bound to oxide or hydroxide is evident in the precipitation of $Al(OH)_3$ by hydroxide from aqueous solutions of aluminium(III), the solubility of $Al(OH)_3$ in excess OH^-(aq), the instability of the non-existent $Al_2(CO_3)_3$, the acidity of $[Al(H_2O)_6]^{3+}$(aq), the hydrolysis of $AlCl_3$ in moist air, and the protective oxide film on the metal.

3 Aluminium sulphate solution can be made by dissolving bauxite in sulphuric acid; it finds application in the water industry. The highly charged positive ions coagulate negatively charged suspended matter, and carry it down with the precipitate of $Al(OH)_3$, formed by the reaction with HCO_3^- ions present in natural waters.

4 The structures and melting temperatures of the aluminium trihalides show signs of increasing covalent character from three-dimensional AlF_3, through the layer structure of $AlCl_3$, to molecular, Al_2X_6 structures in $AlBr_3$ and AlI_3.

5 With the exception of the metal and its alloys, aluminium occurs only in oxidation number +3 at normal temperatures: the monohalides, formed by heating the trihalides with the metal, disproportionate on cooling.

6 The basic character of oxides/hydroxides increases down Group III, changing from acidic at boron, to amphoteric at aluminium and gallium, to basic at indium and thallium.

7 The ionization energies of gallium, indium and thallium are raised relative to those of the lower elements of Groups I and II by the build up of nuclear charge as d and f shells are filled in the preceding elements. This makes the elements more electronegative, and, thermodynamically, less readily oxidized to oxidation number +3. It also leads to the inert pair effect: a gradual increase in the stability of oxidation number +1 towards the bottom of the Group. Thus, GaCl, GaBr, GaI, InCl, InBr and InI have all been made, and $Tl^+(aq)$ is formed when the metal dissolves in acids.

8 Tl^+ and K^+ have similar sizes; they both form soluble carbonates and hydroxides. Thallium(I) compounds are very poisonous.

SAQ 6 Both gallium and indium metals dissolve in dilute HCl, but only one of them in dilute NaOH. Which one? Explain your reasoning, and write an equation for the reaction with the alkali.

SAQ 7 Which of the two sulphates, Na_2SO_4 and $Al_2(SO_4)_3$, will decompose more readily on heating to give a metal oxide and oxides of sulphur?

SAQ 8 How can ionic radii and/or electronegativities be used to explain the structural changes in the solid aluminium halides from AlF_3 to AlI_3?

SAQ 9 Which element, and which oxidation numbers should be most stabilized by the inert pair effect in each of Groups IV and V of the Periodic Table?

4 THE GROUP IV ELEMENTS

4.1 STRUCTURES AND PROPERTIES OF THE ELEMENTS

Group IV (Figure 30) is the second of the Groups we shall be studying in this Block. Like Block 3, it includes a non-metal (carbon), semi-metals (silcon and germanium), and metals (tin and lead). Electronic configurations of the Group IV elements are shown in Table 4.

Carbon forms strong single bonds with itself and is also able to form multiple bonds via pπ–pπ bonding. Both appear in the allotropes of carbon.

Table 4 Electronic configurations of Group IV atoms

Atom	Electronic configuration
C	$[He]2s^22p^2$
Si	$[Ne]3s^23p^2$
Ge	$[Ar]3d^{10}4s^24p^2$
Sn	$[Kr]4d^{10}5s^25p^2$
Pb	$[Xe]4f^{14}5d^{10}6s^26p^2$

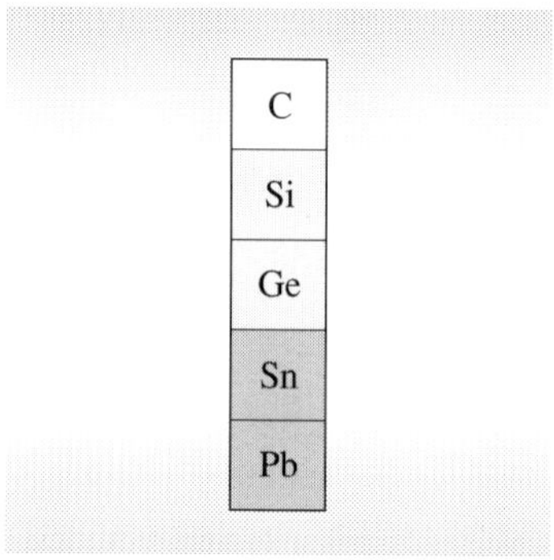

Figure 30 The Group IV elements: metals are shaded dark grey; semi-metals are shaded light grey; the non-metal carbon is unshaded.

■ What are the common allotropes of carbon?

□ Graphite is the stable form of carbon under ambient conditions, whereas diamond is metastable under these conditions (Figures 31 and 32). The newly discovered allotrope of carbon is crystalline C_{60}, buckminsterfullerene (Figure 33). These structures were discussed in detail in Block 2, pp. 47–49.

Silicon and germanium also have the diamond crystal structure. Tin (symbol Sn, from the Latin *stannum*) is polymorphic. Grey tin, a semi-metal with the diamond structure, is the stable form below the transition temperature of 13 °C. Above this temperature, white tin, a metal, is the stable form. The story goes that as Napoleon's army retreated from Moscow in 1812, the tin buttons on their great coats crumbled as they changed from white to grey tin in the extremely cold conditions! French winters, in which the temperature does not fall to Russian levels, would not have affected the buttons,

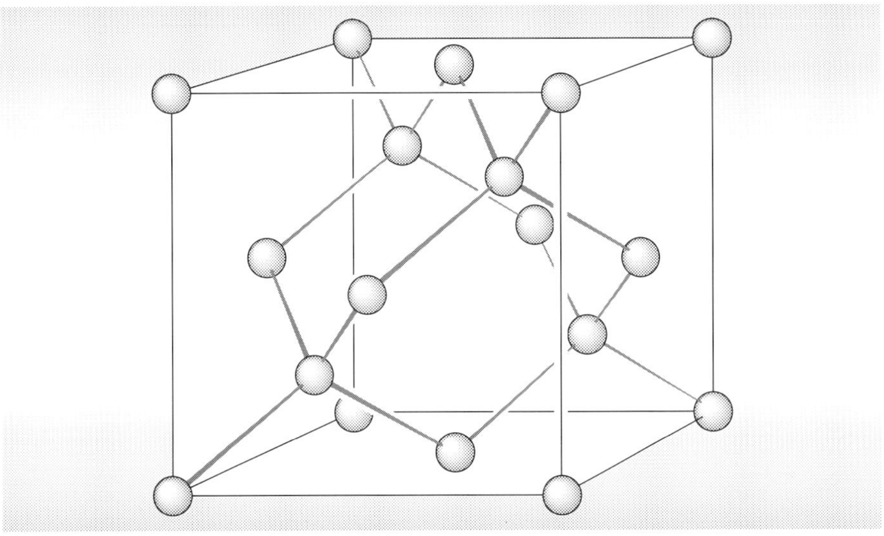

Figure 31 A unit cell of the diamond structure.

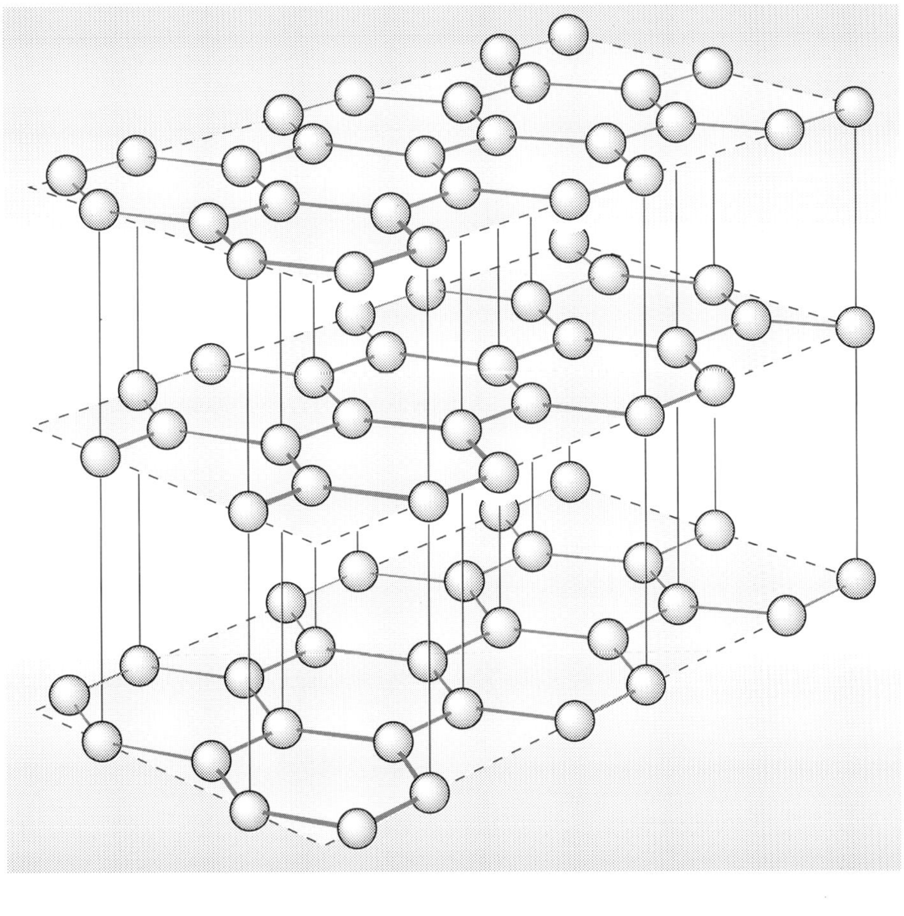

Figure 32 The crystal structure of graphite.

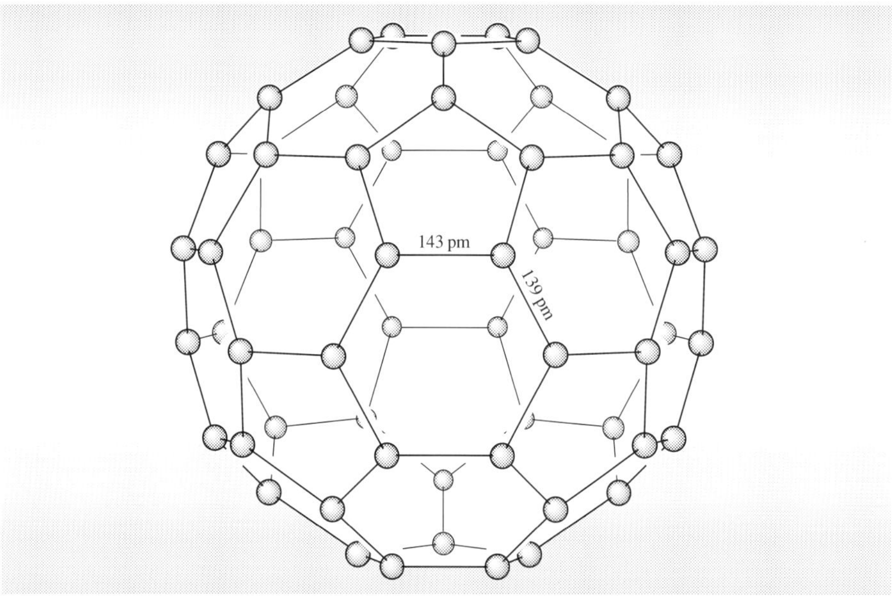

Figure 33 The structure of buckminsterfullerene, C_{60}.

because white tin persists below 13 °C in a metastable state. Tin is also the principal constituent of solder. During Scott's expedition to the South Pole, the petrol cans were found to leak: it is thought that the very low temperatures in the Antarctic caused the tin to change phase and the solder to disintegrate. In white tin, each Sn atom is approximately six coordinate, and four other tin atoms are only a little further away. The last member of Group IV is lead (symbol, Pb, from the Latin *plumbum*), and it has a typical close-packed metal structure.

■ How does the coordination number for the form of the elements stable at room temperature vary down Group IV?

□

graphite		silicon		germanium		white tin		lead
3	→	4	→	4	→	6	→	12

Notice that the structures of the Group IV elements gradually change down the Group as the elements become more metallic in character; from the giant covalent structures of graphite and the **isostructural** (same crystal structures) diamond, silicon, germanium and grey tin, to the metallic structures of white tin and lead. The elements become more metallic as we descend the Group.

As we might expect, the electrical properties of the isostructural forms show a gradation down the Group. Diamond is an insulator, crystalline silicon and germanium, which have metallic lustres, are intrinsic semiconductors, and grey tin conducts almost as well as some metals.

■ What can we say about the band gaps for these elements?

□ The band gaps of the isostructural forms of the elements *decrease* down the Group.

White tin (metallic) is a much better conductor than grey tin. In the plane of the layers, graphite conducts electricity better than either silicon or germanium.

Inspection of Section 9 of the *Data Book* shows you that the homonuclear bond enthalpy terms decrease down the Group: C—C, 347 kJ mol^{-1}; Si—Si, 226 kJ mol^{-1}; Ge—Ge, 188 kJ mol^{-1}; Sn—Sn, 152 kJ mol^{-1}. There is also a fall in melting temperature down the Group: 4 100 °C for diamond; 1 420 °C for silicon; 945 °C for germanium; grey tin changes to white tin before melting (white tin melts at 232 °C); lead melts at 327 °C. The isostructural forms show diminishing hardness down the Group: diamond is the hardest known substance, silicon and germanium are somewhat softer and grey tin is a powder. On Moh's hardness scale, diamond registers 10, silicon 7 and grey tin 1.5.

4.2 CARBON

Carbon

I am an atom of carbon
And carbon is the key
I am the element of life
And you owe yours to me
I am the glue of the Universe
The fixative
used by the Great Model-maker
I play a waiting game
Lie low that's my secret
Take a breath every millenium
but though set in my way
Don't be misled. I'm not inert
I will go down in cosmic history
as an adventurer
For when I do make a move
Things happen and fast
I am an atom of carbon
And carbon is the key
I am the element of life
And you owe yours to me
When the tune is called
I carry the message
to the piper
Take the lead
in the decorous dance
of life and death
Patient, single minded and stable
I keep my talents hidden
Bide my time
Until by time am bidden.

Roger McGough

With four valence electrons and middling electronegativity, carbon forms mainly covalent compounds. The structures of the principal carbon allotropes — diamond, graphite and buckminsterfullerene, C_{60} — are shown in Figures 31, 32 and 33.

When carbon is in its standard thermodynamic state, it has the graphite structure (Figure 32); this is **anisotropic** (the structure is not the same in each principal direction), consisting of sheets of regular hexagons of carbon, like benzene rings fused together. The bond distance in the hexagons (142 pm) is intermediate between the values observed for normal single and double bonds (154 pm in ethane and 134 pm in ethene). The distance between the sheets (335 pm), however, is about double the van der Waals radius of carbon. Graphite, therefore, has less than two-thirds the density of diamond, so that, although graphite is the stable form at normal pressures, it can be turned into diamond at high pressures and temperatures (more than 105 atmospheres at 2 500 °C).

We can account for the bond distance in the graphite sheet in terms of the number of σ and π electrons.

■ Of the four valence electrons, how many are used in σ bonding?

□ Three, to form an electron-pair bond with each of the three neighbours.

This leaves one π electron per carbon atom (as in ethene or benzene). Imagine an infinite π orbital covering the whole sheet, formed from the $2p_z$ atomic orbitals, one on each carbon (perpendicular to the sheets). But since there are three bonds to carbon in the graphite sheet, the π bond order in each is about one-third. The total bond order is thus about $1\frac{1}{3}$, intermediate between a single and a double bond. Graphite is black, conducts electricity in the hexagonal planes, and is used as a lubricant and as the 'lead' in pencils. Its high melting temperature, 3 570 °C, makes it useful as a crucible material for metal casting. Graphite fibres are used to strengthen plastics, finding application in the frames of tennis racquets, for example.

The diamond structure is **isotropic** (the same in all principal directions), with tetrahedrally coordinated carbon. The carbon–carbon distance is what we would expect for single bonds (154 pm). Diamond is transparent, an electrical insulator, refractory, the hardest substance known, and is used in cutting tools.

Both graphite and diamond have extended covalent structures (also called giant molecular structures), and in their chemical reactions they behave similarly; they are unreactive, infusible and insoluble because the giant molecules are held together by strong covalent bonds. Their very different structures, however, can explain their rather different physical properties. In graphite, only van der Waals forces hold the layers together, and they are relatively far apart. This explains graphite's relatively low density and anisotropy. The crystals shear parallel to the planes with ease. However, contrary to a common misconception, this is not the reason why graphite makes a useful lubricant. The lubricant properties of graphite are dependent on the presence of adsorbed layers of water vapour or oxygen: if these are lost due to low pressure or high temperature, then the lubricant properties disappear. For high-vacuum conditions, such as outer space, surface additives are incorporated in order to preserve the low-friction, low-wear properties. Due to the mobile electrons in the giant π orbital, graphite is electrically conducting in the layers, like a two-dimensional metal, but it is a poor conductor perpendicular to the layers. Diamond's isotropic three-dimensional structure explains its hardness; it is an electrical insulator, but interestingly it has the highest thermal conductivity of any known substance (approximately five times that of copper), which is why diamond cutting tools used for drilling do not overheat. There is current (1994) research into methods of depositing diamond in thin layers so that it can be used as a protective layer on microchips, in order to conduct away the heat generated in the circuits, but not interfere with the electrical performance.

In 1985 Robert Curl, Harry Kroto, Richard Smalley and their students, discovered new forms of pure carbon now known as the **fullerenes**. The most famous of these molecules is C_{60}, **buckminsterfullerene**, with the shape of a soccer ball (Figure 33 and Case Study 3, Figure 4). However, it was not until five years later that Wolfgang Kratschmer and Donald Huffman found a method of making fullerenes in large enough quantities to confirm the structural predictions of the earlier discoveries: they are made by passing an electric arc between two graphite electrodes in a partial atmosphere of helium. The fullerenes produced in this way consist mostly of the two most stable forms, C_{60} (75%) and C_{70} (23%), together with a few higher molecular mass ones. Smaller fullerenes can also be made, down to C_{32}. They can be dissolved in benzene and recrystallized. Buckminsterfullerene (a truncated icosahedron) is the most symmetrical of the molecules; it consists of twelve pentagons joined to twenty hexagons, and all the carbon atoms in the structure are equivalent. The carbon–carbon distances between adjacent hexagons are 139 pm. The carbon–carbon linkages shared by hexagons and pentagons are 143 pm, similar to the carbon–carbon distance in the graphite hexagons of 142 pm. We can think of these structures as a carbon sheet with graphite-type delocalized bonding, which bends back on itself to form a polyhedron. The fullerene molecules all have twelve pentagons of carbon atoms linking together different numbers of hexagons: C_{70} has 25 hexagons, and its shape has been likened to that of a rugby ball (Figure 34).

The C_{60} buckyballs (as they are often known) close pack together in the crystals, forming a face-centred cubic array of balls about 100 pm apart. Many experiments are currently taking place on these molecules. It has been found that fullerenes can be formed around metal atoms such as potassium, caesium, lanthanum and uranium. A potassium/C_{60} salt, potassium buckide, K_3C_{60}, has also been formed (Figure 35). It is a metallic crystal with a cubic close-packed array of buckyballs, in which potassium ions occupy both the octahedral and the tetrahedral holes. K_3C_{60} becomes a superconductor (a substance with no electrical resistance) below 18 K. If rubidium is substituted for potassium, the superconductivity is observed up to 30 K. There are many possibilities in the chemistry of these molecules yet to be explored.

Other forms of carbon that you may have come across are *charcoal* and *soot*. These are forms of microcrystalline carbon. **Activated charcoal**, made by heating charcoal in steam at high temperatures to remove impurities, is very porous and highly adsorbent* . It is commonly used in the food industry to decolorize and purify foodstuffs, and is one of the constituents of water filters, where it helps to remove the heavy-metal contaminants.

* A substance is said to be *adsorbed* on a surface when there is bonding between the molecule and the surface; this may be weak van der Waals bonding or stronger covalent bonding.

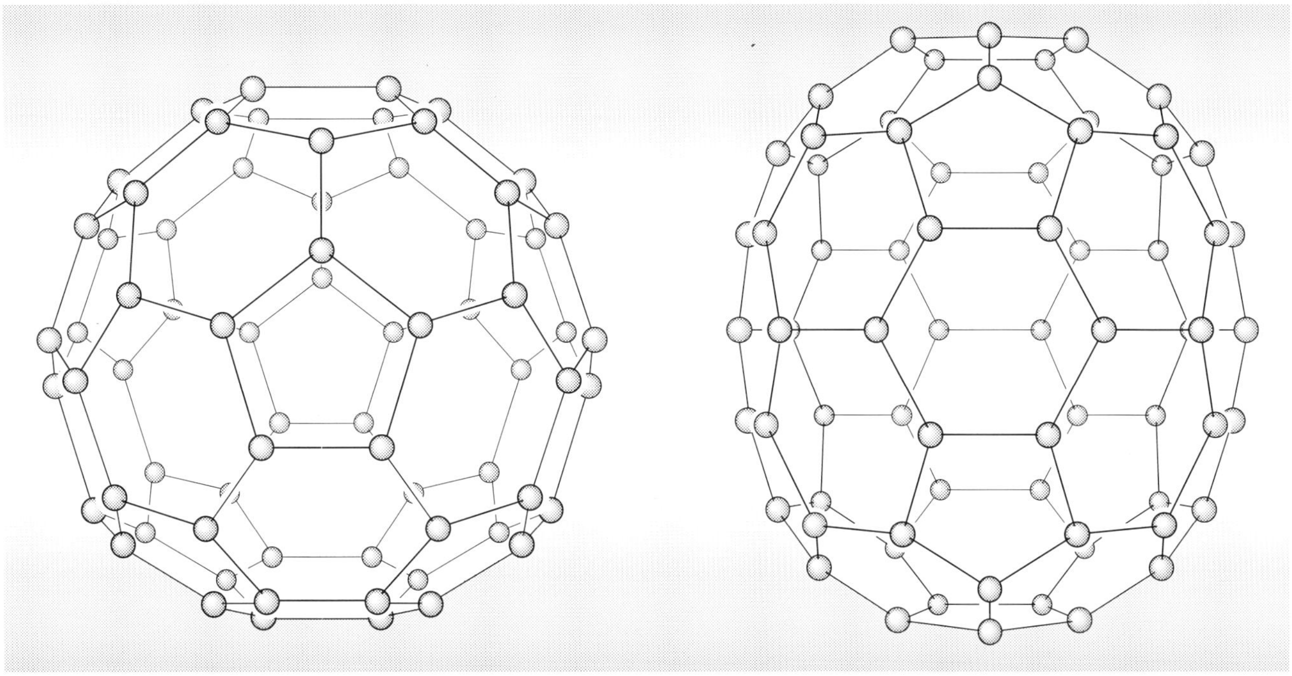

Figure 34 The structure of C_{70}, seen from two different angles.

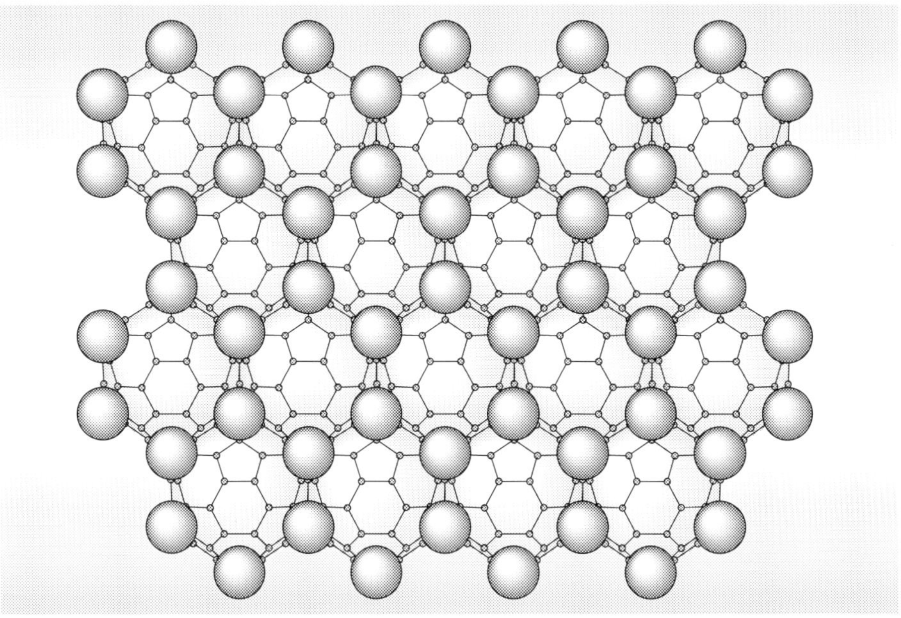

Figure 35 The structure of potassium buckide, K_3C_{60}.

4.2.1 CARBIDES — MOLECULAR, SALT-LIKE AND INTERSTITIAL

Carbides are compounds in which carbon is the more electronegative partner. We shall come across silicon carbide, SiC, again later when we look at the chemistry of silicon; this is a very hard material, manufactured in large quantities as an abrasive known as carborundum. Its hardness can be accounted for by the fact that its crystal structure is the same as that of diamond, with alternate carbon atoms replaced by silicon (Figure 36). It is also used to manufacture high-strength, heat-resistant ceramics for use in high-performance car engines (Plate 1).

Salt-like carbides, such as Be_2C, are sometimes called methanides since they produce methane on hydrolysis. Ethynides (formerly known as as *acetylides*) are also salt-like carbides, and contain the $^-C{\equiv}C^-$ ions. On hydrolysis they produce ethyne, C_2H_2 (former name *acetylene*); this was a reaction that was of utility in illumination before the general availability of electric lighting (and is still sometimes used for arc lights), because ethyne burns with a bright flame. They are formed by the alkali and alkaline earth metals, an example being calcium ethynide (commonly known as *calcium carbide*), CaC_2.

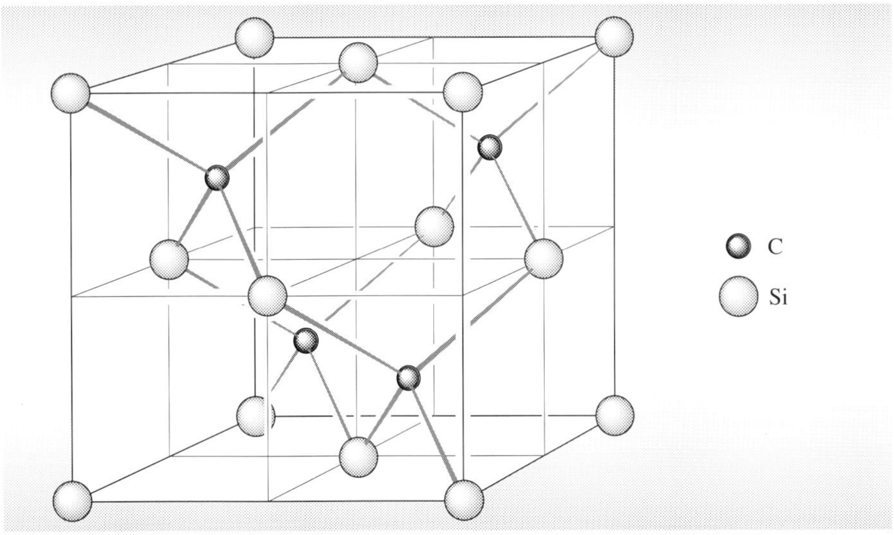

Figure 36 The structure of SiC.

Interstitial carbides are so called because some of the octahedral and tetrahedral holes in a close-packed metal structure may be occupied by 'guest' atoms; that is, the carbon atoms are in interstices rather than lattice sites. Some of these materials have great industrial importance; steel is the best-known example. The interstitial atoms are primarily carbon, but there are also some boron, nitrogen, silicon, phosphorus and other atoms. Steels, which contain 0.5–1.5% of carbon, look and behave like metals, but are harder, higher melting and more corrosion resistant than pure iron. Many transition elements form such non-stoichiometric carbides. Other transition metals, such as tungsten and titanium, also form interstitial carbides with a much higher proportion of carbon, which renders these materials very hard. This enables them to be used in cutting tools.

4.2.2 OXIDES OF CARBON

The common oxides of carbon are carbon monoxide, CO, and carbon dioxide, CO_2. Carbon monoxide is a flammable gas, which burns with a blue flame; it is formed when carbon is burnt in a limited supply of oxygen; in an excess of oxygen, carbon dioxide forms. Carbon monoxide is colourless, odourless and highly poisonous. Its toxicity accounted for the use of coal gas (the forerunner of natural gas) as a ready route to suicide, since CO was one of the principal constituents. The gas is almost non-polar and is one of very few neutral oxides. It has a very short interatomic distance (Structure **14**) and high molar bond enthalpy (1 076 kJ mol^{-1}); the strong bond is due to the π overlap of the 2p orbitals (Block 4, Section 7.4).

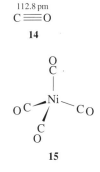

14

Carbon monoxide very readily forms coordination compounds, particularly with transition metals in a low oxidation number. One important industrial use of these compounds is found in the **Mond process** for purifying nickel (developed by L. Mond in 1899). The impure metal is reacted with carbon monoxide to give the volatile **carbonyl compound** nickel tetracarbonyl, $Ni(CO)_4$ (Structure **15**). The $Ni(CO)_4$ vapour is subsequently decomposed to give pure nickel.

15

The toxicity of carbon monoxide is a consequence of its strong coordinating properties. Part of our respiration cycle involves the coordination, and subsequent release, of oxygen, O_2, to the haemoglobin in the blood. The oxygen molecule coordinates to an iron atom in the centre of the haemoglobin molecule (Figure 37a).

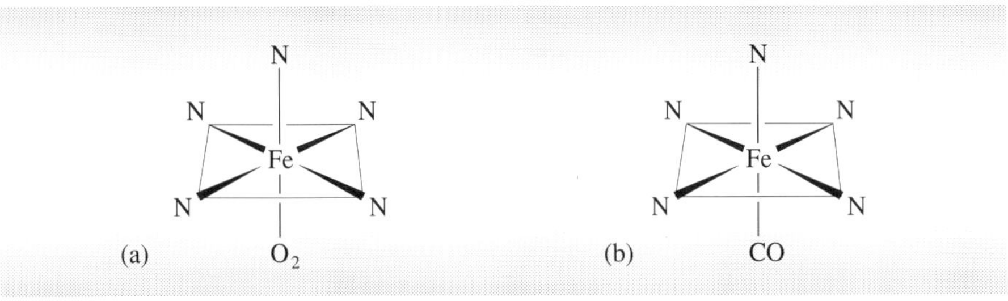

Figure 37 (a) The arrangement of atoms around iron in oxyhaemoglobin (the nitrogen atoms are part of a large organic molecule); (b) arrangement of the atoms around iron in carboxyhaemoglobin.

Carbon monoxide is able to coordinate preferentially with haemoglobin iron (Figure 37b), thus preventing coordination with oxygen. The Fe—CO coordination is very strong and is essentially irreversible, so even low concentrations of CO can soon put all the haemoglobin in the blood out of action: oxygen is prevented from reaching the tissues, leading to suffocation. It is a particularly insidious poison because of its lack of smell. Accidents can happen when, for example, gas heaters are badly ventilated; inhalation of the fumes can induce unconsciousness without the victim ever being aware of a problem. In modern times it has become a health risk, particularly for people with heart disease, both from the high levels of CO pollution created by car engines and from tobacco smoking.

SAQ 10 *(Revising Block 3)* Describe what you understand by a metal complex or coordination compound. Does $Ni(CO)_4$ fit your description?

SAQ 11 *(Revising Block 6)* Would you expect to observe a microwave and/or an infrared spectrum for CO?

SAQ 12 *(Revising Block 4)* What shape does VSEPR theory predict for the CO_2 molecule? What is the bond order?

SAQ 13 *(Revising Block 6)* How many normal modes of vibration do you predict for CO_2? Will they be active in the infrared or Raman or both?

SAQ 14 *(Revising Blocks 4 and 6)* Would you expect the stretching frequency of the carbon–oxygen bond in CO to be higher or lower than for a carbon–oxygen bond in CO_2? How would isotopic substitution to form ^{13}CO affect the stretching frequency?

The chemistry of carbon dioxide is very different from that of carbon monoxide. First, let's revise what you may already know about it. Carbon dioxide is a colourless, odourless gas, which is heavier than air. It can be made in the laboratory by dropping strong acid on to a carbonate (see Block 3, Equations 35 and 36):

$$CaCO_3(s) + 2H^+(aq) = Ca^{2+}(aq) + H_2O(l) + CO_2(g) \quad (24)$$

The experimental set-up for this preparation is illustrated in Figure 38.

As noted in SAQ 12, CO_2 is a linear molecule containing two C=O bonds (Structure **16**). When CO_2 is cooled to −78 °C at normal pressure, it turns straight into a solid without going through a liquid phase. This solid is known as **dry ice**; it can be used to keep food frozen and for some medical procedures such as the freezing of verrucas. It is also commonly used to produce mist effects in theatrical productions: it cools the air below the dew point*, which leads to the formation of clouds of water vapour. The reverse process, the conversion of a solid compound, like solid CO_2, directly into its gaseous form, is known as **sublimation**.

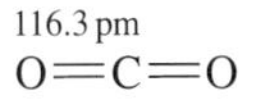

16

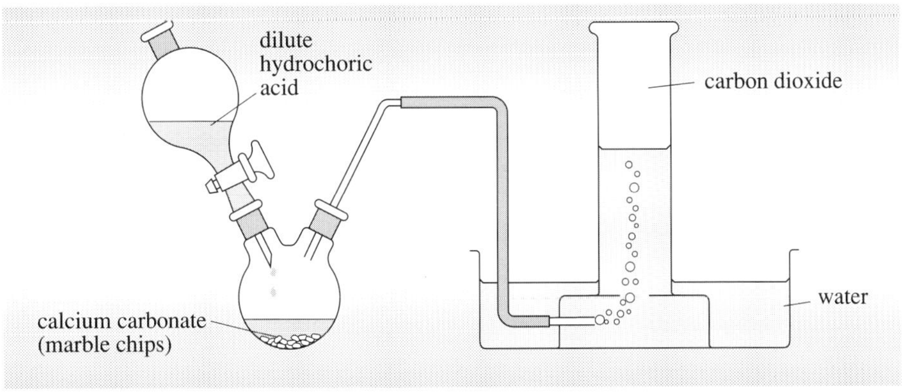

Figure 38 The usual laboratory preparation of CO_2 is to add dilute hydrochloric acid to calcium carbonate (in the form of marble chips).

* Dew point is the temperature at which the air becomes saturated with water vapour; further cooling then leads to the deposition of water as dew on the ground or mist in the air.

In general, CO_2 does not support combustion, so it will extinguish a lighted splint. This property finds application in the use of CO_2 in fire extingushers. An exception is very reactive metals. Thus, magnesium burns in CO_2 to form the oxide and soot:

$$2Mg(s) + CO_2(g) = 2MgO(s) + C(s) \qquad (25)$$

Carbon dioxide is only slightly soluble in water; there is partial reaction to form a weakly acidic solution containing the hydrogen carbonate ion, HCO_3^-:

$$CO_2(aq) + H_2O(l) = HCO_3^-(aq) + H^+(aq) \qquad (26)$$

You may see references to 'carbonic acid', H_2CO_3. However, there is no evidence for the existence of such a molecular species; as Equation 26 indicates, the hydrogen carbonate ion is in equilibrium with hydrated CO_2, not carbonic acid.

CO_2 is bubbled into soft drinks under pressure to make them fizzy; soda water tastes only faintly acid because only a small proportion of the dissolved gas is present as hydrogen carbonate ions: most of it is CO_2 hydrogen-bonded to water, and the equilibrium in Equation 26 lies well to the left. But this equilibrium means that rain falling through even unpolluted skies will be slightly acidic (we pick up on this point again in Case Study 4).

The presence of CO_2 is detected by the **lime water test**. Lime water is a solution of calcium hydroxide in water; CO_2 initially reacts with it to form a fine white precipitate of calcium carbonate, which makes it look milky:

$$Ca(OH)_2(aq) + CO_2(g) = CaCO_3(s) + H_2O(l) \qquad (27)$$

If more CO_2 is bubbled through, the precipitate disappears due to the formation of the hydrogen carbonate ion:

$$CaCO_3(s) + CO_2(g) + H_2O(l) = Ca^{2+}(aq) + 2HCO_3^-(aq) \qquad (28)$$

The atmosphere contains about 0.03% by volume of CO_2, and this is involved in several cycles. CO_2 is produced by respiration of plants and animals, but is used by plants in the photosynthesis of carbohydrates. The fossilization of plants in the Carboniferous Period (285–360 million years ago) was responsible for the production of coal.

Marine organisms use a form of calcium carbonate in their shells. Some of these organisms live in coral reefs, but others swim about in the sea, and when they die their shells fall to the sea bed. These shells eventually become sedimentary rocks such as limestone, which may subsequently be metamorphosed into marble.

The CO_2 dissolved in groundwater plays an important part in the weathering of both carbonate and silicate rocks. Insoluble carbonates such as chalk or limestone are slowly taken into solution as soluble hydrogen carbonates. This occurs because an 'insoluble' carbonate such as $CaCO_3$ is in equilibrium in water with a very small concentration of its ions:

$$CaCO_3(s) = Ca^{2+}(aq) + CO_3^{2-}(aq) \qquad (29)$$

■ What will happen if one of these ions is removed, for instance by reaction with hydrogen ions derived from Equation 26:

$$CO_3^{2-}(aq) + H^+(aq) = HCO_3^-(aq) \qquad (30)$$

□ The equilibrium is disturbed, and more will dissolve to take its place (Le Chatelier's principle).

Silicate rocks are very slowly decomposed or weathered. The alkali metals are removed as soluble carbonates and alkaline earth metals as soluble hydrogen carbonates. Some $SiO_2(aq)$ is also removed in solution, leaving a solid residue of $SiO_2(s)$ as quartz, and clays (which are aluminosilicates). (The chemistry of carbonates is also discussed in some detail in Block 3, Section 12.3; we pick up on it again in Case Study 4.)

CO_2 is produced in the respiration (and fermentation and decay) of animals and plants. The pH of blood is influenced by the concentration of dissolved CO_2 (produced by the oxidation of carbohydrates, etc., in the tissues), and this controls the activity of the lungs.

CO_2 is also produced by the combustion of fuels. The CO_2 content of the atmosphere is thought to have increased by some 10% in the last 150 years, due to an increasing population, increased use of fossil fuels, and also due to the clearing of forests, which has decreased the involvement of CO_2 in photosynthesis. There is international concern that this increased concentration of CO_2 will contribute to 'global warming', because of the **greenhouse effect,** carbon dioxide being one of the main greenhouse gases. The visible and ultraviolet radiation reaching us from the Sun is absorbed by the Earth, but re-radiated in the *infrared* part of the spectrum. As we saw in SAQ 13, carbon dioxide absorbs infrared radiation, thus effectively trapping this energy in the atmosphere. Scientists are concerned that an increase in this trapped energy could cause a partial melting of the polar ice-caps, and thus raise the levels of the oceans. How serious a danger this poses is currently a matter of great debate.

■ Why don't the main atmospheric gases, N_2 and O_2, contribute to the greenhouse effect?

□ Nitrogen and oxygen are homonuclear diatomic molecules, which do not have a permanent dipole moment; this does not change as they vibrate. As this is the condition for absorbing infrared radiation (Block 6, Section 3.1), they do not trap infrared energy radiated from the Earth's surface.

4.2.3 SUMMARY OF SECTION 4.2

1 Carbon forms three main allotropes. Diamond is the hardest known substance; it contains tetrahedral carbon covalently bound to four other carbon atoms in an infinite array. Graphite contains flat layers of carbon hexagons; the 2p orbitals perpendicular to the layers interact to form an extended π orbital. Graphite conducts electricity in the plane of the layers. There are newly discovered forms of carbon known as fullerenes; buckminsterfullerene, C_{60}, has a truncated icosahedral structure, featuring linked pentagons and hexagons of carbon.

2 Carbon forms binary carbides in which it is the more electronegative element. SiC is a hard abrasive substance with the same atomic arrangement as diamond. The salt-like metal carbides such as Be_2C and CaC_2 form methane and ethyne respectively, on hydrolysis. Interstitial carbon atoms in metals form alloys such as steel; the carbon has a profound effect on the physical properties of the metal.

3 Carbon, as a second-row element, has the ability to form multiple bonds by the π overlap of its 2p orbitals.

4 Carbon monoxide, CO, is a poisonous gas, important because of its ability to form coordination compounds: the formation of the volatile carbonyl $Ni(CO)_4$ is used in the purification of nickel.

5 Carbon dioxide, CO_2, is produced naturally by respiration. It forms a weakly acidic solution with water, and is used industrially for putting the 'fizz' in soft drinks.

4.3 SILICON

Silicon is the second most abundant element in the Earth's crust (26% by mass).

■ What is the most abundant element?

□ Oxygen; this has an abundance of 46% by mass.

Silicon always occurs in combination with oxygen and other elements, mostly as silica, SiO_2, or silicates. Silica is converted to many useful silicon-containing products, as shown in Figure 39.

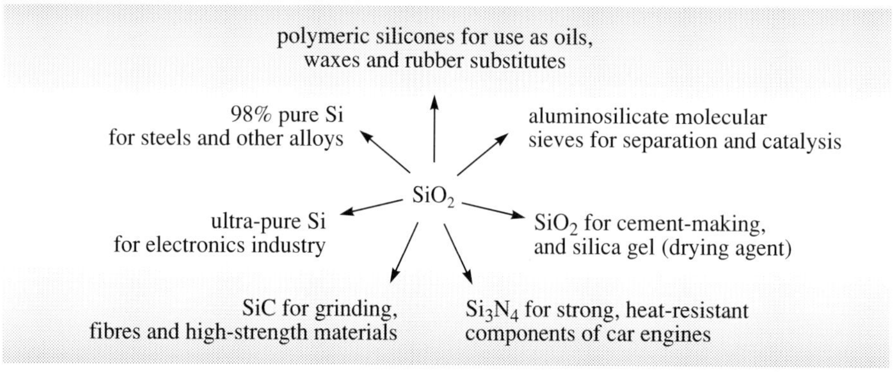

Figure 39 Uses of silicon and its compounds.

Reduction of silica with carbon yields elemental silicon. This is carried out on a large scale in huge electric arc furnaces, and the product is 98–99% pure silicon.

SAQ 15 *(Revising Block 1)* Figure 40 shows the Ellingham diagram for SiO_2 and CO. At what temperature does carbon reduce SiO_2? Calculate the temperature T at which $\Delta G^{\ominus}_{m}$ for Reaction 31 becomes zero, ignoring the phase change Si(s) = Si(l), which occurs at 1 683 K. Use any values you need from the *Data Book.*

$$\tfrac{1}{2}SiO_2(s) + C(s) = CO(g) + \tfrac{1}{2}Si(s) \qquad (31)$$

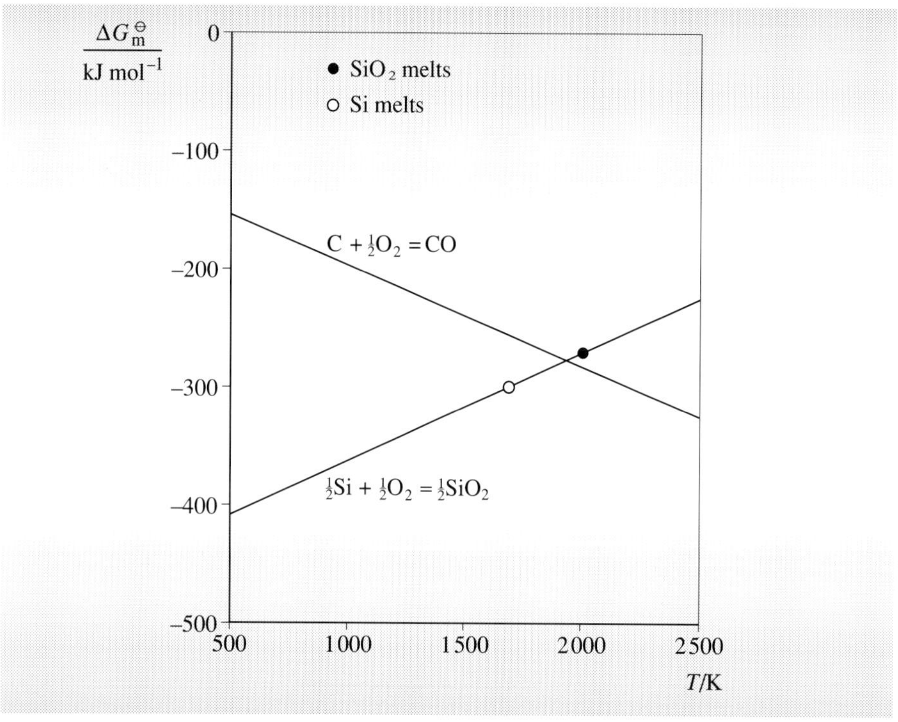

Figure 40 Ellingham diagram for SiO_2 and CO.

Unpurified silicon (98% pure) is suitable for the formation of silicon-based chemicals such as silicones and silicon alloys. Silicon can form compounds and alloys with many metals. These materials, known as **silicides**, have a wide range of stoichiometries and properties, ranging from metallic to covalent. In particular, silicon is an important minor component of many types of steel and of aluminium alloys such as are employed in light-alloy car engines.

Silicon (and germanium) is an intrinsic semiconductor (Case Study 2). It is required in an extremely pure form by the electronics industry, with levels of key impurities, such as phosphorus and boron, of 1 atom in 10^{10} atoms of Si! High-purity silicon is also used to produce high-purity silica for optical fibres. To make silicon this pure,

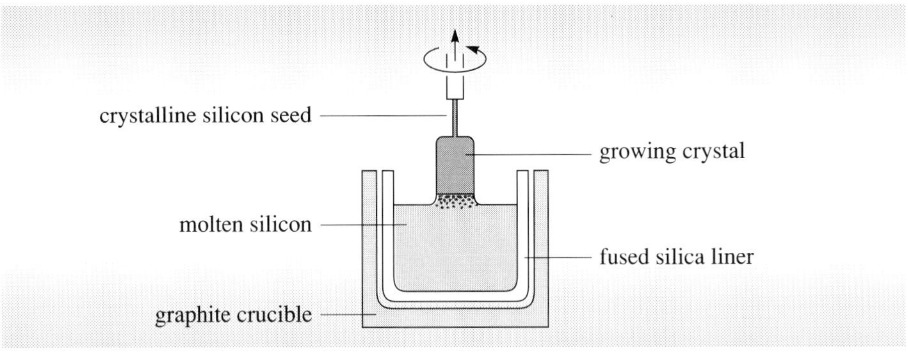

Figure 41 The Czochralski process for producing very pure silicon, in which a single crystal is pulled from the melt.

unpurified silicon is reacted with HCl in a fluidized bed reactor at 300 °C to convert it to the very volatile compound trichlorosilane, $SiHCl_3$:

$$Si(s) + 3HCl(g) = SiHCl_3(g) + H_2(g) \quad (32)$$

The $SiHCl_3$ is then very carefully distilled and finally decomposed again on rods of high-purity silicon at 1 000 °C to give **polycrystalline silicon** (many small crystals). For the electronics industry this material is converted to large rods, which are made into single crystals in two different ways. One method involves melting the silicon in an argon atmosphere (Figure 41), dipping a single-crystal rod into the melt as a seed, and slowly withdrawing an ever-lengthening single crystal (the **Czochralski process**). In the other method a polycrystalline rod of silicon, to which a single-crystal seed is attached at one end, is held vertically (Figure 42), and a molten zone is caused to traverse its length, starting at the seed end (the **float-zone process**). The final single-crystal rods of silicon are ground into cylinders and cut in thin slices, which, after grinding, etching and polishing, are called **silicon wafers**. These metallic looking wafers are processed by the electronics industry into intricate semiconductor devices. Plate 2 shows a rod of single-crystal silicon.

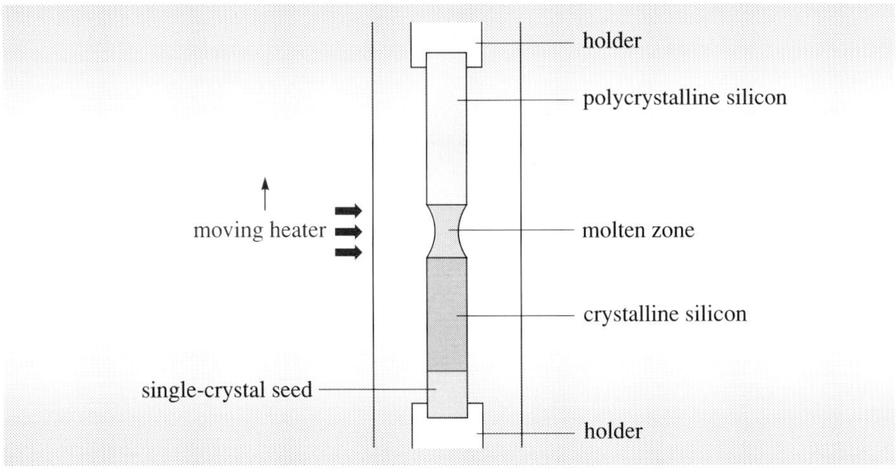

Figure 42 Float-zone method for the purification of silicon.

4.3.1 BONDING IN SILICON COMPOUNDS

■ What is the electronic configuration of the Si atom?

□ $[Ne]3s^23p^2$

Silicon in its compounds is usually four coordinate. Unlike carbon, however, it can increase its coordination number to five or six, if electrons are donated by atoms of another molecule or ion, thus forming a Lewis acid–Lewis base complex: for instance, salts of the complex ions SiF_5^- and SiF_6^{2-} are known.

SAQ 16 *(Revising Block 4)* Using VSEPR theory, predict the shapes of the SiF_5^- and SiF_6^{2-} anions, and deduce their symmetry point groups.

In SiF_5^- and SiF_6^{2-} we can see again the idea that not all compounds obey the 'stable octet' rule. It is common for elements of the third row to increase their coordination number to five or six (and higher in later rows), and we shall meet more examples in the chemistry of phosphorus and sulphur.

Compounds of silicon in oxidation number +2, of general type SiX_2, are called **silylenes**; they are mostly unstable at room temperature, and are important as reaction intermediates.

■ What would you predict to be the shape of the silylenes?

□ VSEPR theory predicts that the two bonding pairs of electrons and the lone pair will give bent molecules. Indeed this is observed for the structures that have been determined.

Table 5 compares some bond enthalpy terms for single bonds in carbon compounds and in silicon compounds.

Table 5 Bond enthalpy terms for C and Si

Bond	B/kJ mol^{-1}	Bond	B/kJ mol^{-1}
C—C	347	Si—Si	226
C—H	413	Si—H	318
C—O	358	Si—O	466
C—F	467	Si—F	597
C—Cl	346	Si—Cl	400
C—Br	290	Si—Br	330
C=O	770	Si=O	638
C=C	612	Si—C	307
C≡C	838		

Note that the Si—Si and Si—H bond enthalpies are significantly lower than those of C—C and C—H. However, with the more electronegative elements from Groups VI and VII, the silicon bond enthalpies are *higher* than the carbon bond energies: the silicon bonds are more polarized than the carbon bonds because silicon is less electronegative than carbon, and the attraction between the differently polarized ends of the bonds leads to extra strength of the silicon bonds. As we saw in Block 5, Section 7.3, Pauling used this additional bonding energy, which he called the *ionic resonance energy*, to calculate values for the electronegativities of the elements. We can also use his equation

$$B(\text{A—B}) - \tfrac{1}{2}[B(\text{A—A}) + B(\text{B—B})] = C(\chi_A - \chi_B)^2 \qquad (33)$$

(Block 5, Equation 43) in the opposite sense, to obtain an estimate of the extra bond strength that we expect from the difference in electronegativity. The electronegativities of silicon and fluorine are 1.8 and 4.0, respectively, and so the ionic resonance energy is given by

$$C(\chi_F - \chi_{Si})^2 = 96.5(4.0 - 1.8)^2 = 467\,\text{kJ mol}^{-1}$$

If we compare the value for the bond enthalpy term of Si—F in Table 5 with the average of the Si—Si and F—F values (*Data Book*), the measured additional bond strength is 405 kJ mol^{-1}. (The discrepancy arises from the fact that the values derive from an empirical equation and the electronegativity values are averaged from many data.)

You will know from your work at the CHEM 999 Summer School that carbon tetrachloride is a rather pungent liquid that is immiscible with water and does not react with it. By contrast, silicon tetrachloride is a reactive liquid, which immediately fumes in contact with air, producing white clouds of hydrochloric acid.

SAQ 17 *(Revising Block 1)* Using values from the *Data Book*, calculate $\Delta G_m^\ominus$ at 298.15 K for the two hydrolysis reactions below, and predict which reaction is thermodynamically more favourable.

$$CCl_4(l) + 2H_2O(l) = CO_2(g) + 4HCl(g) \quad (34)$$

$$SiCl_4(l) + 2H_2O(l) = SiO_2(s) + 4HCl(g) \quad (35)$$

As your calculations should show, both Reactions 34 and 35 are favourable on thermodynamic grounds, so, clearly, the stability of CCl_4 with respect to hydrolysis must be due to the fact that its hydrolysis is immeasurably slow; that is, it is a kinetic effect and is due to the mechanism of the reaction.

So, a comparison of bond enthalpies between carbon and silicon by itself is a poor guide to relative chemical reactivity. This is because many carbon compounds of the type CX_4 are extremely inert to reactions that would require an expansion of the octet around carbon, whereas, as implied at the start of this Section, many SiX_4 compounds can readily increase their coordination number by adding an electron pair donor like H_2O. This gives the opportunity for subsequent reactions, such as loss of HX and formation of SiO_2. The reactivity of silicon compounds accounts for the fact that silicon is only found in nature combined with oxygen, whereas carbon is found in kinetically stable compounds with hydrogen, nitrogen, sulphur and chlorine, as well as with oxygen.

4.3.2 SILICON–OXYGEN COMPOUNDS

We have already noted (Section 2) the great difference in physical properties between CO_2 and SiO_2, oxides of elements in the same Group of the Periodic Table: one is a colourless gas, the other a hard crystalline solid. In CO_2 the $2p_x$ and $2p_y$ orbitals on C and O overlap to give pπ–pπ bonding; the result is a discrete molecule containing strong C=O double bonds, O=C=O (see Table 5). In SiO_2 the greater strength of the Si—O single bond makes it thermodynamically preferable for Si to form Si—O—Si bridges rather than Si=O double bonds. This leads to the crystalline structure of silica (Figure 43), in which silicon is tetrahedrally coordinated by oxygen, and each oxygen atom is shared with a neighbouring tetrahedron (see Block 2, Section 6).

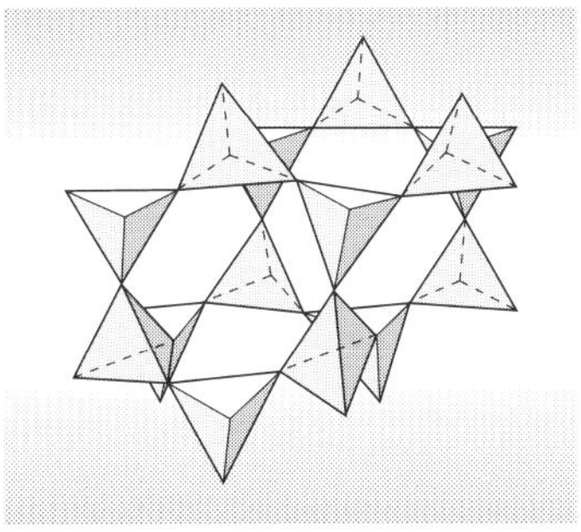

Figure 43 Schematic representation of cristobalite, one of the crystalline forms of silica, which is based on tetrahedral SiO_4 groups joined at the apices. Quartz, the more common form of of silica, has a similar, but more condensed, structure.

Silica, SiO_2, occurs in several crystalline forms, quartz (or rock crystal; see Plate 3) and cristobalite being the most common. Quartz is one of the commonest minerals in the Earth, occurring as sand on the seashore, as a constituent of granite and flint, and, in less pure form, as agate and opal. Silica behaves like an acid in that it combines on heating with the basic oxides and carbonates of the metals to form silicates, which constitute most of the important rock-forming minerals and their weathering products, clays and soils. A great part of the Earth's crust consists of silicates.

Many thousands of silicates are known, formed by the linking of SiO_4^{4-} tetrahedra via oxygen sharing. The classification of these compounds is of considerable importance to geologists. Before the advent of X-ray crystallography, mineralogists were faced with an intractable problem when they tried to classify the silicates according to the empirical formulae that they obtained by chemical analysis. In those days the formula types identified were the orthosilicates, M_4SiO_4 (where M is a univalent metal), metasilicates, M_2SiO_3, pyrosilicates, $M_6Si_2O_7$, and many more such categories. But many minerals that were well characterized (by their crystalline form, for example) would not fit into any of these chemical categories, and when chemists tried to isolate the supposed silicic acids (in which the metal cations have been replaced by hydrogen), they ran into difficulties.

After the X-ray crystallographic studies of the silicates by W. H. and W. L. Bragg, and L. Pauling in the 1920s, the minerals could be classified by structure (Figure 44).

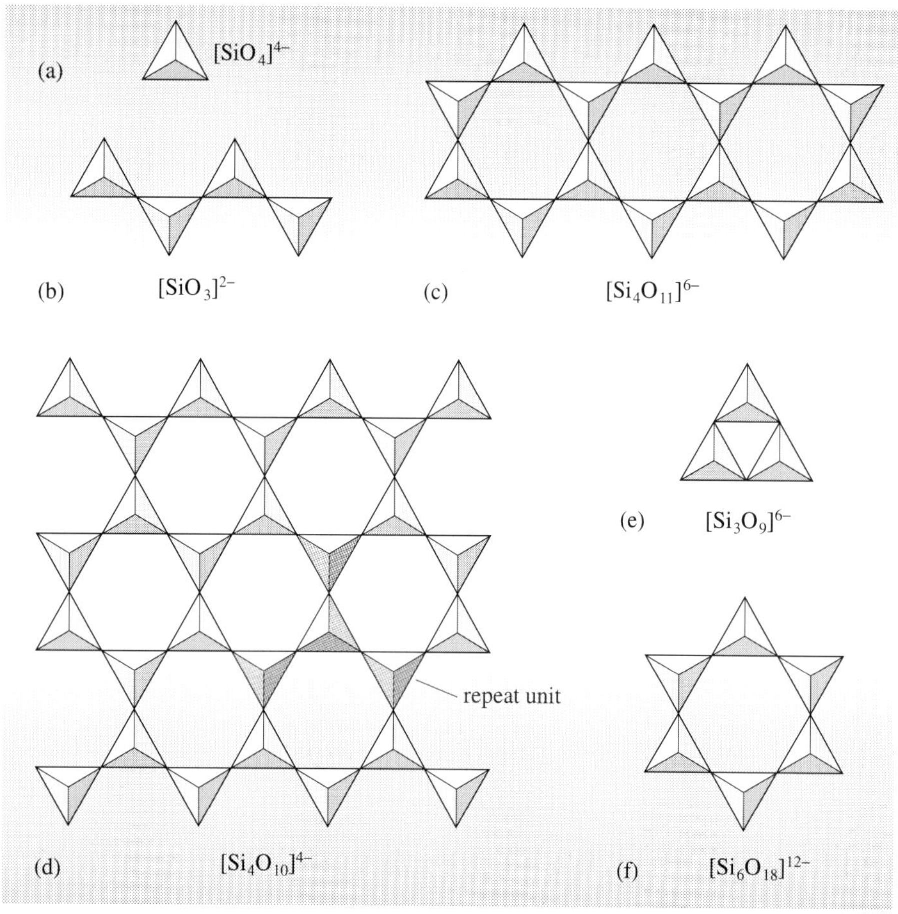

Figure 44 A structural classification of mineral silicates: (a) the $[SiO_4]^{4-}$ unit; (b) chains; (c) double chains; (d) infinite layers; (e) ring of three tetrahedra; (f) ring of six tetrahedra.

Clearly, the sharing of oxygens between the SiO_4^{4-} tetrahedra to give polyanions give rise to the various empirical formulae; the **isomorphous replacement*** of one metal by another metal was responsible for many of the difficulties in the classification of minerals by chemical composition. The demonstration also that aluminium in a silicate crystal could be either six coordinate or four coordinate (replacing silicon), even in the same structure, solved many puzzles in understanding the silicates.

The silicate structures are most conveniently discussed in terms of the $[SiO_4]^{4-}$ unit (Figure 44a), in which there is tetrahedral coordination of silicon by oxygen. Some minerals, for example olivines and garnets, contain $[SiO_4]^{4-}$ anions as discrete tetrahedra. These can be described as salts of orthosilicic acid, $Si(OH)_4$ or H_4SiO_4, which is a very weak acid.

In most silicates, however, the $[SiO_4]^{4-}$ tetrahedra are linked by oxygen sharing (Structure **17**). Here two $[SiO_4]^{4-}$ link together to form a different anion.

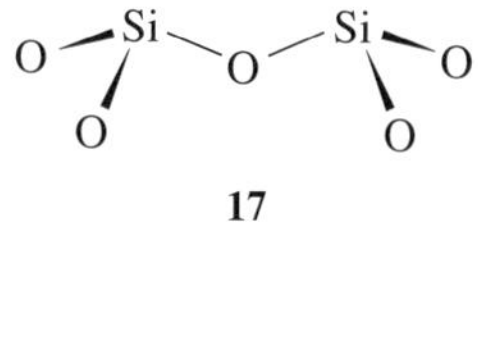

17

■ What is the charge *n* on this anion?

□ There is one negative charge on each singly bonded oxygen, and the shared oxygen is neutral. The anion is thus $[Si_2O_7]^{6-}$; in other words, $n = 6$.

* Replacement of one metal by another while retaining the same crystal structure.

Since the charges can readily be worked out in this way, they are often omitted from diagrams of the structures, such as those shown in Figure 44.

By sharing one or more oxygen atoms, the tetrahedra are able to link up to form chains, rings, layers, etc. Some examples are discussed below.

1 *Discrete* $[SiO_4]^{4-}$ *units* Examples are Ca_2SiO_4, found in mortars and Portland cement, and zircon, $ZrSiO_4$ (Figure 44a).

2 *Chains* $[SiO_4]^{4-}$ units share two corners to form infinite chains (Figure 44b). The repeat unit is $[SiO_3]^{2-}$. Minerals with this structure are called **pyroxenes**, for example $CaMg(SiO_3)_2$ and $MgSiO_3$. The silicate chains lie parallel to one another, and are linked together by the cations that lie between them.

3 *Double chains* Here alternate tetrahedra share two and three oxygen atoms, respectively (Figure 44c). This class of minerals, which is known as the **amphiboles**, contains most asbestos materials. The repeat unit is $[Si_4O_{11}]^{6-}$.

4 *Infinite layers* All the tetrahedra share three oxygen atoms (Figure 44d). The repeat unit is $[Si_4O_{10}]^{4-}$. Examples are: talc, $Mg_3(OH)_2Si_4O_{10}$ (which contains a sandwich of two layers, with octahedrally coordinated Mg^{2+} between the layers); micas, such as $KMg_3(OH)_2Si_3AlO_{10}$; clay minerals, such as kaolinite and bentonite.

5 *Rings* Each $[SiO_4]^{4-}$ unit shares two corners as in the chains. Figures 42e and 42f show three and six tetrahedra linked together; rings also may be made from four tetrahedra. An example of a six-tetrahedra ring is beryl (better known as emerald) $Be_3Al_2Si_6O_{18}$ (see Plate 4). The rings lie parallel with metal ions between them.

6 *Three-dimensional structures* $[SiO_4]^{4-}$ tetrahedra share all four oxygens, giving the same structure as in silica, SiO_2. However, if some of the silicon atoms are replaced by the similarly sized atoms of the Group III element aluminium (that is, if $[SiO_4]^{4-}$ is replaced by $[AlO_4]^{5-}$), then other cations must be introduced to balance the charges. Such minerals include the feldspars, the most abundant of the rock-forming minerals; the zeolites, which are used as ion exchangers, water softeners and molecular sieves (Plate 5); the ultramarines, which are coloured silicates manufactured for use as pigments (lapis lazuli, shown in Plate 6, is a naturally occurring mineral of this type).

Here again, you should be able to see an approximate correlation between the solid-state structures and the physical properties of the compound. We have provided some samples of silicates in the Home Experiment Kit to illustrate this point. (Unfortunately the budget wouldn't allow us to include an emerald!)

ACTIVITY Now see if you understand these concepts by examining the silicate observation samples from the Home Kit. ■

HOME EXPERIMENT

EXAMINATION OF SILICATE SAMPLES

Cement Contains *discrete* $[SiO_4]^{4-}$ units. It is soft and crumbly.

Asbestos Contains double *chains* of $[SiO_4]^{4-}$ units. It is characteristically stringy and fibrous.

Mica Contains infinite *layers* of $[SiO_4]^{4-}$ units. The weak bonding between layers is easily broken, and accordingly micas show perfect cleavage parallel to the layers.

Granite Contains feldspars (the matrix of orange crystals is potassium feldspar; there is also a small amount of white calcium feldspar), which are based on three-dimensional $[SiO_4]^{4-}$ frameworks, and are therefore much harder.

X-ray crystallographic work on silicates also helped to explain the apparent complexity of the so-called 'silicic acids'. Orthosilicic acid, H_4SiO_4, can be made by the hydrolysis of tetrachlorosilane (old name, silicon tetrachloride).

■ What is the equation for this reaction?

□ $SiCl_4(s) + 4H_2O(l) = H_4SiO_4(aq) + 4HCl(aq)$ (36)

The solution is gelatinous and becomes cloudy on standing. This is because the silicic acid polymerizes by splitting off water, to form Si—O—Si bridges, which results in a mixture of molecules of different chain lengths. This process is discussed at more length in Section 5.5.5.

■ Form a chain molecule (diagrammatically) from four $HO{-}Si(OH)_2{-}OH$ molecules, by splitting off three (4 − 1) molecules of water. What is the repeat unit in this chain and what is the overall equation for the process?

□ The repeat unit is $[SiO_3]^{2-}$ (see Figures 44b and 45). The equation is

$$4H_4SiO_4(aq) = H_{10}Si_4O_{13}(aq) + 3H_2O \quad (37)$$

These long-chain acids are called 'metasilicic acid', and are analogous to the silicate chains found in the pyroxenes.

Ordinary 'sodium silicate' or 'calcium silicate' are, in fact, metasilicates. Sodium silicate may be made by heating sodium carbonate, Na_2CO_3, with pure sand (SiO_2) in a furnace to form a glass. If this is ground up in water, it gives a syrup called 'water glass'. This is used for painting cement (to reduce dust), and was used in the past for preserving eggs.

As we saw in Block 3, Section 12.3, ordinary window glass (soda glass) is made by a continuous process in which sand (SiO_2) is reacted at 1 400 °C with fused carbonates of sodium, potassium and calcium (limestone and some sodium sulphate). At these temperatures they behave as a mixture of the oxide and carbon dioxide gas (for instance, $CaO + CO_2$). The liquid is stirred by the evolution of CO_2, H_2O (from the hydrated salts) and SO_3.

■ Write equations for some of these reactions.

□ $Na_2CO_3 + SiO_2 = Na_2SiO_3 + CO_2$ (38)

$$CaCO_3 = CaO + CO_2 \quad (39)$$

$$CaO + SiO_2 = CaSiO_3 \quad (40)$$

$$Na_2SO_4 + SiO_2 = Na_2SiO_3 + SO_3\text{, etc.} \quad (41)$$

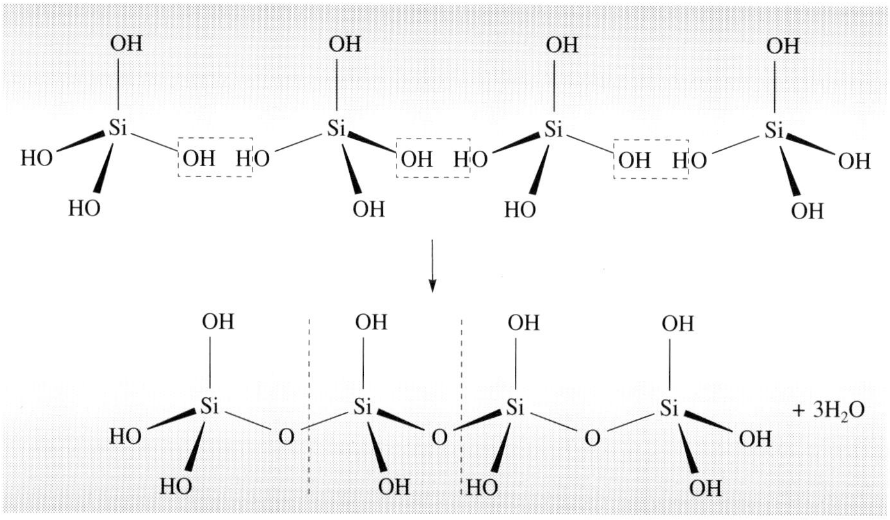

Figure 45 Formation of a metasilicic acid by splitting out water from orthosilicic acid. Coloured broken lines indicate the repeat unit.

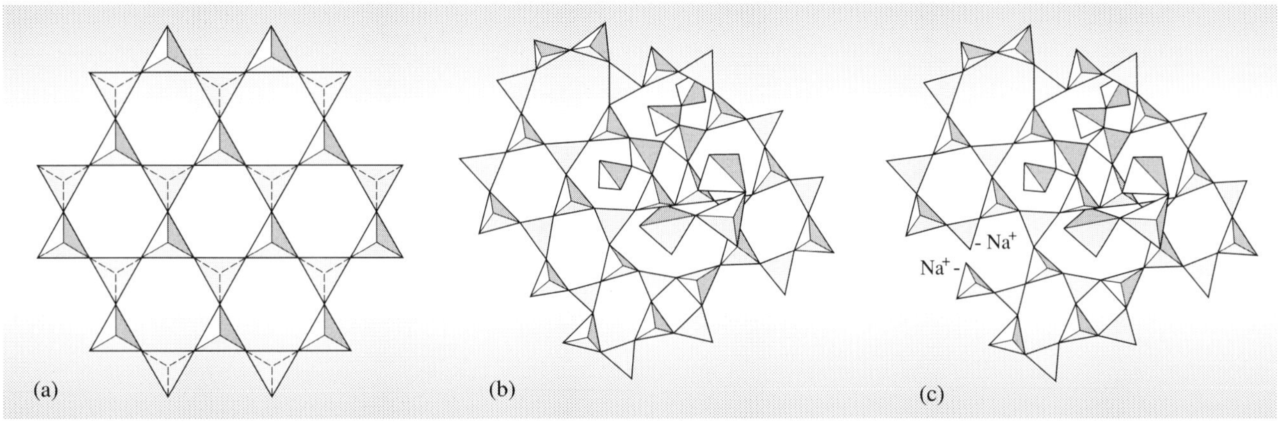

Figure 46 (a) Linked $[SiO_4]^{4-}$ tetrahedra in idealized crystalline SiO_2; (b) a simple SiO_2 glass showing corner-linked, tangled $[SiO_4]^{4-}$ chains; (c) the addition of one molecule of Na_2O, causes an Si—O—Si group to form two $Si—O^-Na^+$ interactions.

At these high temperatures, sodium oxide, for instance, will combine with SiO_2, opening up the Si—O—Si linkages to form two $Si—O^-$ bonds (Figures 46 and 47); the Na^+ ions are coordinated to the O^- sites on the three-dimensional Si—O network (Figure 46c). Because of the high viscosity of the melt, glasses do not form a sufficiently regular array in order to crystallize during cooling. We can visualize a silicate glass structure as tangled chains and branched chains of corner-sharing $[SiO_4]^{4-}$ tetrahedra, with Na^+, K^+ and Ca^{2+} ions between them. The molten glass is drawn off and floated on pure molten tin to form sheets; this gives the glass a flat smooth surface, so that the grinding and the polishing of former times are no longer necessary. This glass is known colloquially as float glass.

Molten SiO_2 itself crystallizes very slowly, and usually solidifies to a glass (Figure 46b). Glassy silica is used for the construction of certain specialized glassware for chemistry laboratories because it is very resistant to chemical attack; it softens and melts at a high melting temperature (1 500 °C), and is transparent to visible, ultraviolet and near-infrared light. Above about 1 100 °C, however, it tends to crystallize (devitrify).

Pyrex glass, used in laboratories and kitchens, etc., expands less than soda glass when heated (this is why it does not crack when its temperature is suddenly changed), and has a higher melting temperature than ordinary soda glass. As we saw in Section 3.1.5, it is a borosilicate glass; it contains less sodium and potassium than ordinary glass, and no lime at all; the triangular BO_3 groups, which are linked through their vertices, are opened up by the metal oxides, replacing the B—O—B groups with $B—O^-$ bonds, so that the BO_3 groups can be linked through zero, one or two vertices.

Some silicate minerals are found as glasses, but most have had plenty of time in which to crystallize, either from magmas or after the various metamorphic processes. Some of the many applications of silicates are summarized in Table 6.

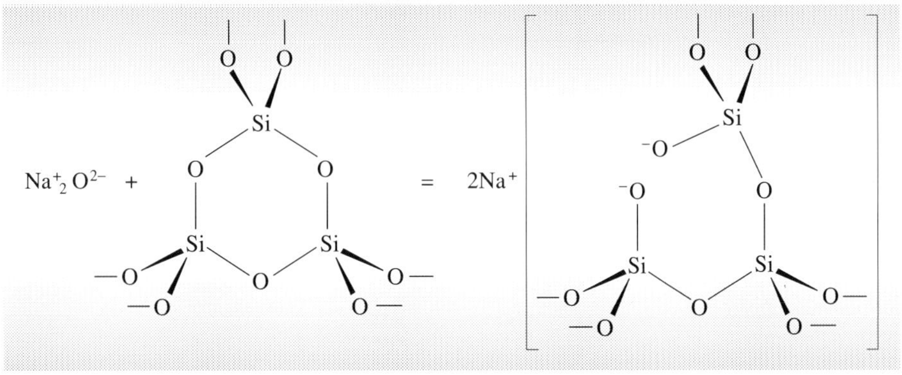

Figure 47 A representation of the structural changes that take place when sodium oxide reacts with silica. Each oxide ion attacks the silica structure, converting a Si—O—Si linkage into two $Si—O^-$ bonds.

Table 6 Uses of silicon–oxygen compounds

Name	Formula	Notes
silica gel	$SiO_2(H_2O)$	amorphous microporous structure with surface area of 700 $m^2\,g^{-1}$; can absorb up to half its weight of water
vitreous silica	SiO_2	clear amorphous silica; high thermal shock resistance; very high softening temperature; transmits u.v. and near i.r. light
pyrogenic or 'fume' silica		made by burning $SiCl_4$ in an oxygen-rich hydrocarbon flame; used for thickening oils and greases
exfoliated vermiculite	$(Mg,Ca)_{0.7}(Mg,Al,Fe^{III})_6[(Al,Si)_8O_{20}](OH)_4.8H_2O$	a natural silicate, which on rapid heating expands to 20 times its original volume; used in packaging and thermal insulation
soda glass	$(SiO_2)_{\approx 7}(Na_2O)_{\approx 2}$ $(CaO)_{\approx 1}(Al_2O_3)_{\approx 0.1}$	90% of all glass made has this composition, or small variants thereof; used in bottles, windows, etc.
Portland cement	$(CaO)_a(Al_2O_3)_b(SiO_2)_c$	made at high temperature from a range of materials like limestone, anhydrite ($CaSO_4$) and clays; variable composition, and made on an enormous scale for construction purposes
soluble silicates	$(Na_2O)(SiO_2)_n$	viscous opalescent colloidal solutions; used in detergents, adhesives and high-temperature binders
organic silicates	$(RO)_4Si$	colourless liquids, very easily hydrolysed; used as paint binders and cross-linking agents; first recognized by Mendeléev in 1860
silicones	$(R_2SiO)_n$	a very wide range of polymers, having —Si—O—Si— chains, with organic substituents on the silicon atoms; fluids, resins, rubbers, etc.

4.3.3 HALOSILANES

Direct reaction of silicon with fluorine, chlorine, bromine or iodine, gives stable but moisture-sensitive compounds, SiX_4. The reaction between F_2 and Si is extremely violent, and SiF_4 is much more conveniently made by reacting concentrated H_2SO_4 with a mixture of SiO_2 and sodium fluoride, NaF (Equation 42). The driving force for this reaction is the greater strength of the Si—F bond compared with Si—O.

$$2H_2SO_4 + 4NaF + SiO_2 = 2Na_2SO_4 + 2H_2O + SiF_4 \qquad (42)$$

Tetrafluorosilane, SiF_4 (old name, silicon tetrafluoride) is a gas at room temperature. The other halosilanes are liquids with boiling temperatures increasing with molecular mass. The reaction of tetrafluorosilane with water is rather different from that of other halosilanes, in that it forms both SiF_6^{2-} ions and SiO_2; the strongly electronegative fluoride ligands bring out the maximum coordination number of six for silicon.

■ Write a balanced equation for the reaction of $SiF_4(g)$ and $H_2O(l)$.

□ $3SiF_4(g) + 2H_2O(l) = 4H^+(aq) + 2SiF_6^{2-}(aq) + SiO_2(s)$ (43)

The other halosilanes do not give SiX_6^{2-} ions; with excess water they react vigorously and completely to give SiO_2 and the hydrogen halide:

$$SiX_4(g) + 2H_2O(l) = SiO_2(s) + 4HX(g) \qquad (44)$$

Halosilanes containing Si—Si bonds are made via reactions of the unstable silicon dihalides, SiX_2, or from metal silicides. Thus, passing chlorine over calcium silicide, Ca_2Si, at 150 °C gives Si_2Cl_6 and higher homologues of the alkane-like series Si_nCl_{2n+2}.

4.3.4 COMPOUNDS OF SILICON WITH HYDROGEN AND ALKYL GROUPS

It has been known since the pioneering work of Alfred Stock in the 1920s, that magnesium silicide, Mg_2Si (made by heating Mg and Si together) reacts with dilute hydrochloric acid to give a series of compounds called **silanes** with the general formula Si_nH_{2n+2}, which are analogous to the alkanes.

■ Write an equation for the formation of the lowest silane.

□ $Mg_2Si(s) + 4H^+(aq) = 2Mg^{2+}(aq) + SiH_4(g)$ (45)

It has been shown that both straight-chain and branched-chain compounds are present, but the silanes are rather reactive, and the series has only been explored for values of *n* up to about 8. The lower members of the series, SiH_4, Si_2H_6 and Si_3H_8, are better made by reduction of the corresponding chlorosilanes, Si_nCl_{2n+2}, with lithium aluminium hydride, $LiAlH_4$.

The silanes tend to be very reactive compounds and are spontaneously flammable in air. They have to be handled using special techniques to isolate them from contact with oxygen; this involves the use of a vacuum system or keeping them under a nitrogen atmosphere. They do not react with pure water or dilute acids, but hydrolyse rapidly in the presence of even a trace of alkali, liberating hydrogen and forming silicates.

SAQ 18 Using values from the *Data Book*, calculate $\Delta G_m^\ominus$ for the reaction of (a) methane, CH_4, and (b) ethane, C_2H_6, with oxygen to give carbon dioxide and water. Also, calculate $\Delta G_m^\ominus$ values for the analogous reactions of (c) monosilane, SiH_4, and (d) disilane, Si_2H_6.

The values you obtain from the calculations in SAQ 18 show you that in oxygen, both alkanes are unstable with respect to CO_2 and H_2O, and both silanes are also unstable with respect to SiO_2 and H_2O. The silanes are spontaneously flammable in air, but the alkanes are not. Methane (natural gas) can, of course, explode in air if it is sparked!

■ How would you explain this difference?

□ The alkanes possess a *kinetic* stability to reaction with O_2 that the silanes do not. The reasons for this must lie in the *mechanism* of the reactions, and are probably concerned with the larger size of silicon and its ability to expand its octet. (The explosive reaction of methane when sparked is due to the production of radical species, such as $CH_3^•$, which initiate a chain reaction.)

When the hydrogen atoms in silanes are replaced by alkyl or aryl groups, the chemical reactivity is much reduced and the thermal stability is increased. Many organosilicon compounds are made industrially by two different catalytic processes named after their inventors. In the first, the **Rochow process**, chloromethane, CH_3Cl, is reacted with silicon granules in the presence of 10% copper and small amounts of oxides of metals like magnesium, calcium or tin at 300 °C. Under the correct conditions, the main organosilicon product is $(CH_3)_2SiCl_2$. $(CH_3)_3SiCl$ and CH_3SiCl_3 are also formed, so it is necessary to separate the products by fractional distillation. Fifty years after the discovery of this process, the reaction mechanism is still not fully elucidated! In the second method, the **Speier hydrosilation process**, $HSiCl_3$ is reacted with an alkene, $RCH{=}CH_2$, in the presence of a platinum catalyst, to give $RCH_2CH_2SiCl_3$. Controlled hydrolysis of organosilicon chlorides gives compounds known as **silicones** or **polysiloxanes**, compounds with a silicon–oxygen backbone (Figure 48) first discovered by F. S. Kipping.

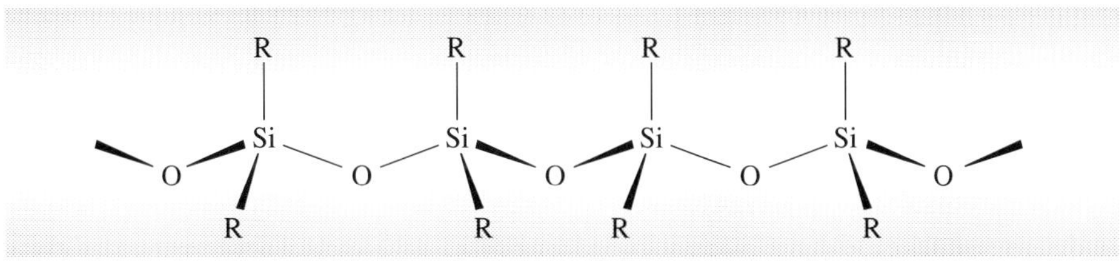

Figure 48 An unbranched silicone chain (R is an alkyl group, usually methyl).

Four different types of structural unit are found in silicones; they are formed by the hydrolyses shown in Equations 46–49:

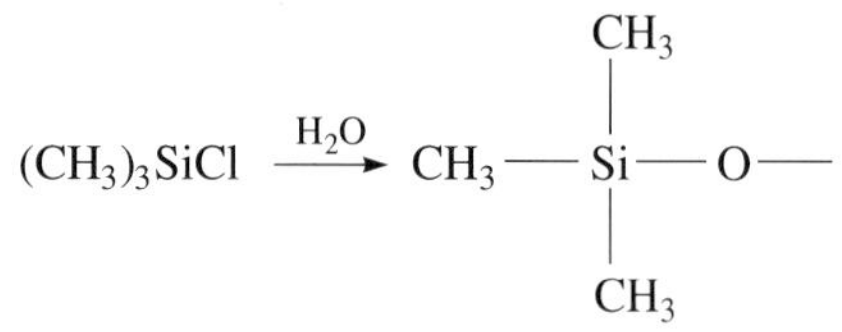

$$(CH_3)_3SiCl \xrightarrow{\text{H}_2\text{O}} CH_3 - \overset{\displaystyle CH_3}{\underset{\displaystyle CH_3}{\overset{\displaystyle |}{\underset{\displaystyle |}{Si}}}} - O - \tag{46}$$

end group

$$(CH_3)_2SiCl_2 \xrightarrow{\text{H}_2\text{O}} -O - \overset{\displaystyle CH_3}{\underset{\displaystyle CH_3}{\overset{\displaystyle |}{\underset{\displaystyle |}{Si}}}} - O - \tag{47}$$

chain group

$$CH_3SiCl_3 \xrightarrow{\text{H}_2\text{O}} -O - \overset{\displaystyle CH_3}{\underset{\displaystyle O}{\overset{\displaystyle |}{\underset{\displaystyle |}{Si}}}} - O - \tag{48}$$

branching group

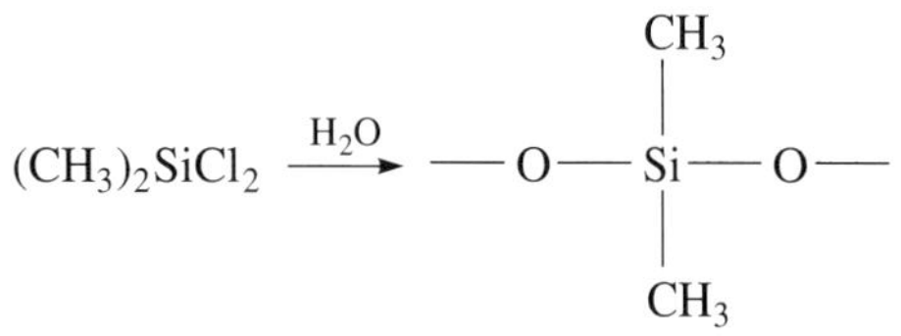

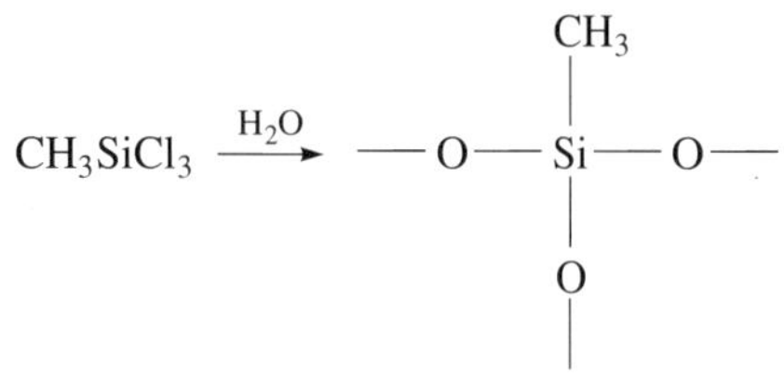

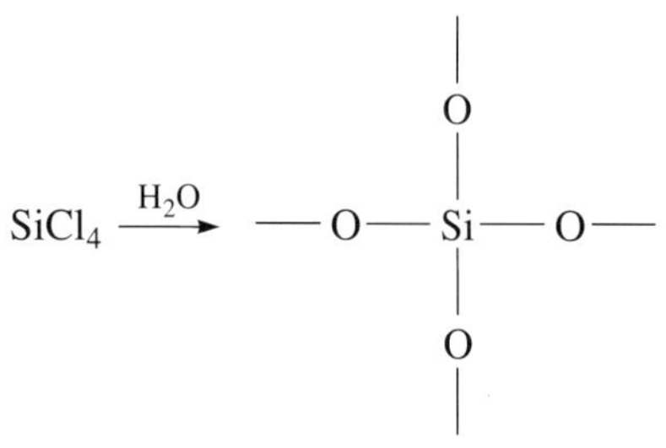

$$SiCl_4 \xrightarrow{\text{H}_2\text{O}} -O - \overset{\displaystyle O}{\underset{\displaystyle O}{\overset{\displaystyle |}{\underset{\displaystyle |}{Si}}}} - O - \tag{49}$$

branching group

The extent of polymerization and branching can be controlled by introducing end-blocking groups such as in Equation 46.

The important feature of the silicones is that they have a backbone containing only the very strong Si—O bonds, giving them good thermal stability. The backbone is surrounded by a sheath of methyl (or other organic) groups. The viscosity of silicones varies little over a wide temperature range. The longer the chain of the silicone, the greater its viscosity, and silicone oils can be tailor-made to a required viscosity. The oils are used for hydraulic fluids, light lubricants, car polishes, cosmetics, etc.

The silicones have low surface tension and are non-toxic; they are used as antifoaming agents in textile dyeing, fermentation, sewage disposal and cooking oils. They also provide an effective remedy for flatulence in both humans and animals by breaking down the emulsion in the stomach.

When the siloxane chain is **cross-linked**, using for instance branching groups such as indicated in Equations 48 and 49, or methylene, $-CH_2-$, groups, the reaction can be used to produce rubber-like materials. These materials are flexible down to very low temperatures, and are used for gaskets, space-suits, soft contact lenses, and inside the body in some spare-part surgery.

A different type of silicone rubber is the so-called 'bouncing' or 'silly putty', which is made by heating a silicone oil with a small amount of boric acid, $B(OH)_3$. The boron is incorporated into the chain, and because of its Lewis acid properties it forms weak inter-chain links, via O—B interactions (Figure 49). The cross-linking causes the

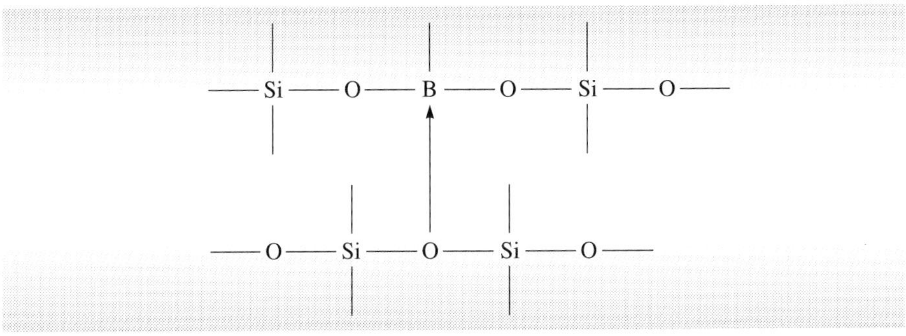

Figure 49 Incorporation of boron in a silicone rubber to give 'bouncing' or 'silly putty'.

material to have the property of a rubber; for instance, it bounces well. On the other hand, if squeezed or stretched, it behaves as a fluid; the chains can slide past one another, and a ball of the material collapses to a flat pool under its own weight. If thrown hard on to a surface, it will also shatter!

Until 1981, compounds containing silicon–silicon double bonds were thought to have only a fleeting existence at room temperature. If they were made, they immediately polymerized to compounds containing silicon–silicon single bonds. Silicon–silicon double bonds are therefore *thermodynamically* unstable with respect to such polymerization. Then the way of stabilizing silicon–silicon double bonds by introducing a *kinetic* barrier, was discovered. If silicon is joined to very bulky groups, polymerization is impossible and the silicon–silicon double bonds exist at ordinary temperatures! The first success was obtained by using the 2,4,6-trimethylphenyl group (known as mesityl, Mes) (Structure **18**); it is bulky enough to prevent the close approach of the double bonds which is needed for polymerization to occur.

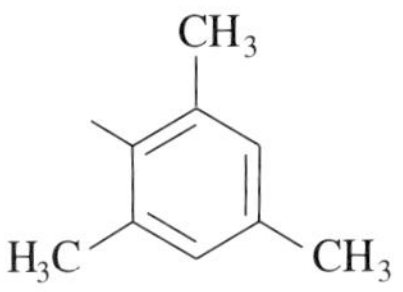

18

The silicon–silicon double-bonded product $(Mes)_2Si{=}Si(Mes)_2$ is known as a **disilylene**. The compound reacts easily with small molecules such as oxygen, which can attack the double bond, but melts without decomposition at 178 °C in the absence of air. A number of similar disilylenes have now been made using different bulky groups. Success has also been achieved in stabilizing silicon–carbon double bonds with bulky groups to prevent the polymerization that otherwise occurs. (We shall also see the same trick used to make phosphorus–phosphorus double-bonded compounds in Section 5.5.6.) It is interesting that nature often makes use of very bulky protein groups to control the chemistry within living cells, such as the reversible O_2 binding of haemoglobin.

The ultimate organosilicon compound is silicon carbide, SiC (see Section 4.2.1). This is a very hard and strong substance, made on a huge scale as a grinding agent by the reduction of SiO_2 with carbon at 2 000 °C:

$$SiO_2(s) + 3C(s) = SiC(s) + 2CO(g) \qquad (50)$$

The crystal structure of silicon carbide, which was shown in Figure 36, is of the zinc blende type. It can be thought of as the diamond structure, in which every alternate carbon atom is replaced by a silicon atom.

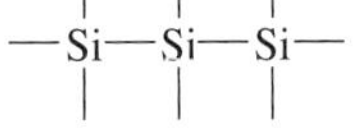

19

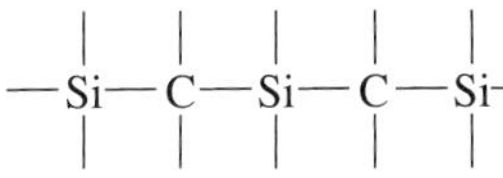

20

Silicon carbide is also made as very strong fibres by the sequence of reactions shown in Equations 51 and 52, in which an infusible polymer containing the silicon backbone shown in Structure **19** is made first. This is converted by heat in an argon atmosphere to an isomer containing the silicon–carbon backbone shown in Structure **20**, which can be melted and spun into fibres. The fibres are heated in air at 300 °C to partly oxidize them and make them infusible, and are then heated in nitrogen to 1 300 °C to convert them to SiC fibres:

$$n(CH_3)_2SiCl_2 + 2nNa = [(CH_3)_2Si]_n + 2nNaCl \qquad (51)$$

$$[(CH_3)_2Si]_n \xrightarrow[\text{argon}]{400\,°C} \left(\overset{\displaystyle H}{\underset{\displaystyle CH_3}{Si}} - CH_2 \right)_n \xrightarrow{\text{spin to fibres by heating in air at 300 °C and then in } N_2 \text{ at 1300 °C}} nSiC \qquad (52)$$

The fibres are used to reinforce aluminium and other materials in aircraft construction and other applications where light weight and high strength are important.

Silicon carbide is resistant to oxidation in air to about 1 600 °C because of the formation of an adherent layer of SiO_2 on its surface.

Silicon nitride, Si_3N_4, made from Si or SiO_2/C mixtures and N_2/H_2 at 1 300–1 600 °C, is also resistant to air oxidation, and is being tested extensively as a light-weight replacement for metal in car and jet engine construction.

4.4 GERMANIUM, TIN AND LEAD

Perhaps the most spectacular feature of these elements is the increasing tendency to form stable divalent species (that is, the occurrence of a +2 oxidation number for the Group IV element in mononuclear compounds). Thus, whereas all the silicon tetrahalides, SiX_4 (where X = F, Cl, Br, I) are known, the mononuclear SiX_2 species (the silylenes) are thermodynamically unstable at room temperature and have been identified as reaction intermediates. By contrast the +2 oxidation number is the most common one for lead, and all the halides PbX_2 are known and stable at room temperature. Of the PbX_4 halides, only PbF_4 is stable at room temperature; $PbCl_4$ decomposes to $PbCl_2$ and chlorine gas above 50 °C, and $PbBr_4$ and PbI_4 are unknown. In the stable dihalides, PbX_2, two of the four valence electrons of lead are exhibiting the inert pair effect, which we first met for the elements at the bottom of Group III (Section 3.3). Although the inert pair of electrons is not used in bonding, its presence in the valence shell usually has an effect on the shape. For instance, the gaseous $SnCl_2$ molecule is observed to have a bond angle of 95°. Another example is provided by the solid-state structures of SnO and PbO (litharge), which have the same crystal structure, containing square-pyramidal units of SnO_4 (or PbO_4), arranged so that the oxygens form parallel layers; the metal atoms lie alternately above and below the oxygen layer, and the lone pair is assumed to lie opposite the metal atom (the O—Sn—O angle here is 75°; see Figure 50).

The inert pair effect is first manifested by germanium, and increases through tin to lead, where we see the dramatic change whereby oxidation number +2 becomes predominant. Clearly, we would only expect the most powerful oxidizing agents such as fluorine (PbF_4) or oxygen (PbO_2) to be able to oxidize lead to oxidation number +4.

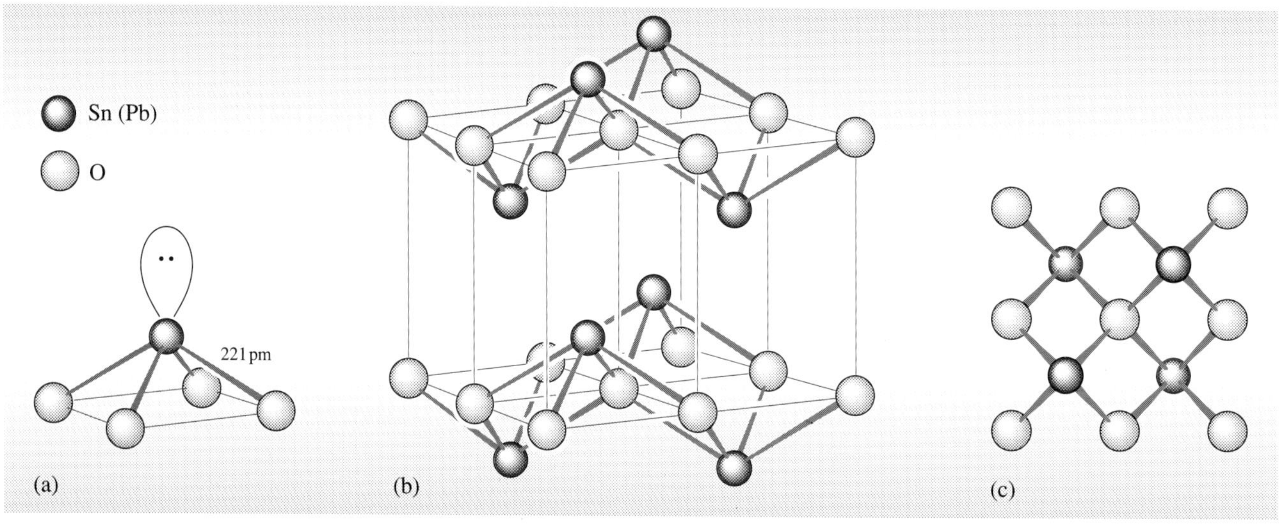

Figure 50 The crystal structure of SnO (and PbO), showing: (a) a square-pyramidal SnO_4 unit; (b) the linking of the square pyramids in layers; (c) looking down on a single layer (the metal atoms above the oxygen layer have heavy-bordered grey circles, and those below the plane have light-bordered grey circles).

SAQ 19 The following hexahalo ions have been characterized for the first four elements of Group IV:

C: none

Si: SiF_6^{2-}

Ge: GeF_6^{2-}, $GeCl_6^{2-}$

Sn: SnF_6^{2-}, $SnCl_6^{2-}$, $SnBr_6^{2-}$, SnI_6^{2-}

What reasons can you think of to explain these differences?

4.5 SUMMARY OF SECTIONS 4.3 AND 4.4

1 Silicon is produced by carbon reduction of SiO_2. It can be purified by forming trichlorosilane, $SiHCl_3$, which is then distilled and decomposed. Single crystals of silicon are made for use in the semiconductor industry.

2 Silicon is usually four coordinate, but as a third-row element it can increase its coordination number to five and six in species such as SiF_5^- and SiF_6^{2-}.

3 The strength of the Si—O bond means that silicon prefers Si—O—Si linkages, unlike carbon, which readily forms C=O double bonds. SiO_2 occurs as the crystalline solid, quartz, with Si tetrahedrally coordinated by four oxygen atoms. SiO_2 combines with basic oxides and carbonates, and so can be said to be acidic in character.

4 Thousands of silicate structures are known. They are formed by the different arrangements of $[SiO_4]^{4-}$ tetrahedra, which link together by bridges through the apical oxygens. Silicon can be replaced in these structures by aluminium to give $[AlO_4]^{5-}$; additional metal cations are then required to balance the charges.

5 Glass is made by reacting sand (SiO_2) with metal carbonates and sodium sulphate. It consists of tangled chains of $[SiO_4]^{4-}$ tetrahedra, with Na^+, K^+ and Ca^{2+} ions in between. Pyrex glass also contains boron.

6 All the binary halides, SiX_4, of silicon are known; they are compounds that hydrolyse rather easily. The hexafluoro ions, MF_6^{2-}, are known for all the Group IV elements except carbon, but the larger iodide ion is only able to form SnI_6^{2-}.

7 Silicon forms a homologous series of very reactive hydrides, Si_nH_{2n+2}, known as the silanes.

8 Hydrolysis of the organosilicon chlorides gives a group of polymeric compounds known as silicones, which contain a —Si—O—Si—O— backbone. These compounds have been developed industrially for many uses such as oils, polishes and rubber-like materials.

9 Recently, compounds have been made containing Si=Si double bonds; polymerization is prevented by the use of bulky substituent groups.

10 The tendency to form divalent compounds increases down the Group; the +2 state is the most common one for lead.

SAQ 20 Why do you think that a homologous series of silicon hydrides, Si_2H_{2n}, analogous to the alkenes, does not exist?

SAQ 21 Use Pauling's equation to predict a value for the ionic resonance energy of a Si—Cl bond, and compare it with the experimental value.

5 THE GROUP V ELEMENTS

5.1 STRUCTURES AND PROPERTIES OF THE ELEMENTS

Group V (Figure 51) is the third of the Groups we shall be studying in this Block. Like Groups III and IV, it includes non-metals (nitrogen and phosphorus), semi-metals (arsenic and antimony), and a metal (bismuth). Electronic configurations of the Group V elements are shown in Table 7.

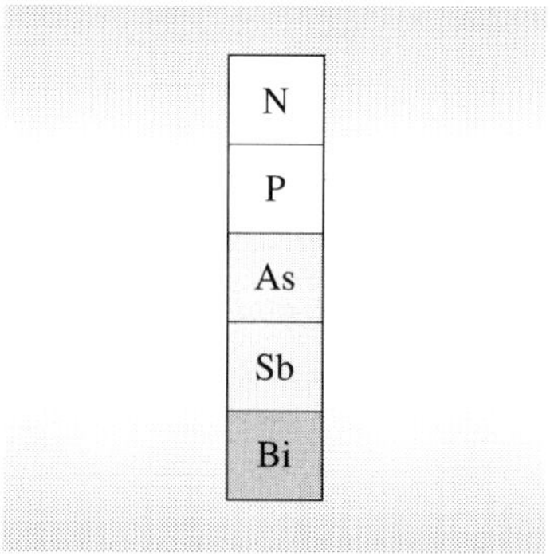

Figure 51 The Group V elements: the metal (bismuth) is shaded dark grey; semi-metals are shaded light grey; non-metals are unshaded.

Nitrogen occurs as the main component — about 80% — of the air around us. The homonuclear diatomic molecule is very unreactive. The bonding in nitrogen is discussed in Block 4, Section 6.2.3: the molecule has a very high bond strength due to the triple bond (the molar bond enthalpy is 945 kJ mol^{-1}) and a very high ionization energy. This chemical inertness means that so-called '**nitrogen fixation**' — the conversion of nitrogen into useful chemical compounds such as ammonia, nitric acid and nitrates — is difficult, requiring a catalyst and a lot of energy. Nitrogen compounds are needed to make fertilizers, plastics and synthetic fibres for instance.

Table 7 Electronic configurations of Group V atoms

Atom	Electronic configuration
N	$[He]2s^22p^3$
P	$[Ne]3s^23p^3$
As	$[Ar]3d^{10}4s^24p^3$
Sb	$[Kr]4d^{10}5s^25p^3$
Bi	$[Xe]4f^{14}5d^{10}6s^26p^3$

Phosphorus occurs in various phosphate minerals, the main one being fluoroapatite, $3Ca_3(PO_4)_2.CaF_2$ (empirical formula $Ca_5(PO_4)_3F$). There are a number of allotropic forms of elemental phosphorus, the most common of which is the toxic white phosphorus, P_4 (also called α-phosphorus), which has the tetrahedral structure shown in Figure 52. White phosphorus has to be stored under water because it oxidizes spontaneously in air, slowly warming up and then bursting into flame.

■ What is the apparent bond angle in the P_4 molecule?

□ The PPP angle is 60°, and this is unusually low.

At very high temperatures, phosphorus dissociates into atoms; at somewhat lower temperatures (1 500 °C), the vapour contains diatomic molecules, P_2, analogous to nitrogen, N_2, whereas below about 800 °C the vapour or liquid consists almost entirely of P_4 molecules. If the vapour is cooled and condensed under water, the tetrahedral white phosphorus shown in Figure 52 is formed.

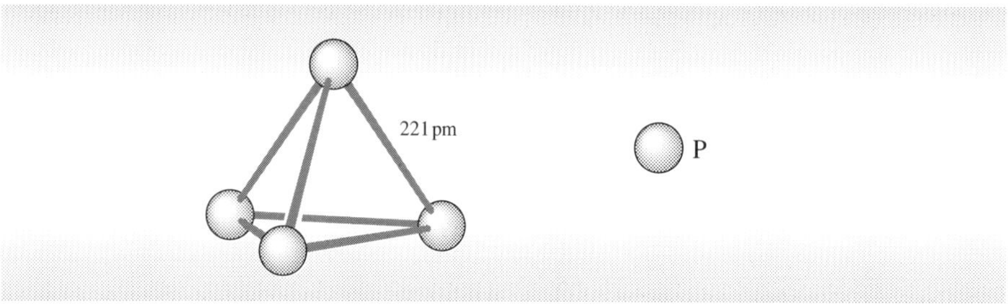

Figure 52 The structure of white phosphorus, P_4.

Elemental white phosphorus can be converted to the less dangerous, red, violet or black forms (Figure 53) by heating under pressure. On heating, the P_4 molecules open out and join up to form chains, two-dimensional layers and three-dimensional structures of covalently bonded phosphorus atoms. Several of the allotropes, such as red phosphorus, are **amorphous**; that is, they have no long-range order of atomic packing as is found in crystals. However, as illustrated in Figure 53, some of the varieties of phosphorus do have a well-defined crystal structure. The polymeric allotropes of phosphorus are insoluble, non-flammable and non-toxic, unlike P_4.

■ What do you notice about the coordination number of phosphorus in all these structures?

□ In each case, phosphorus forms three single bonds; that is, it has a coordination number of three.

Arsenic, As, antimony, Sb, and bismuth, Bi, have a metallic appearance but are brittle. They are isostructural, with layer structures (Figure 54).

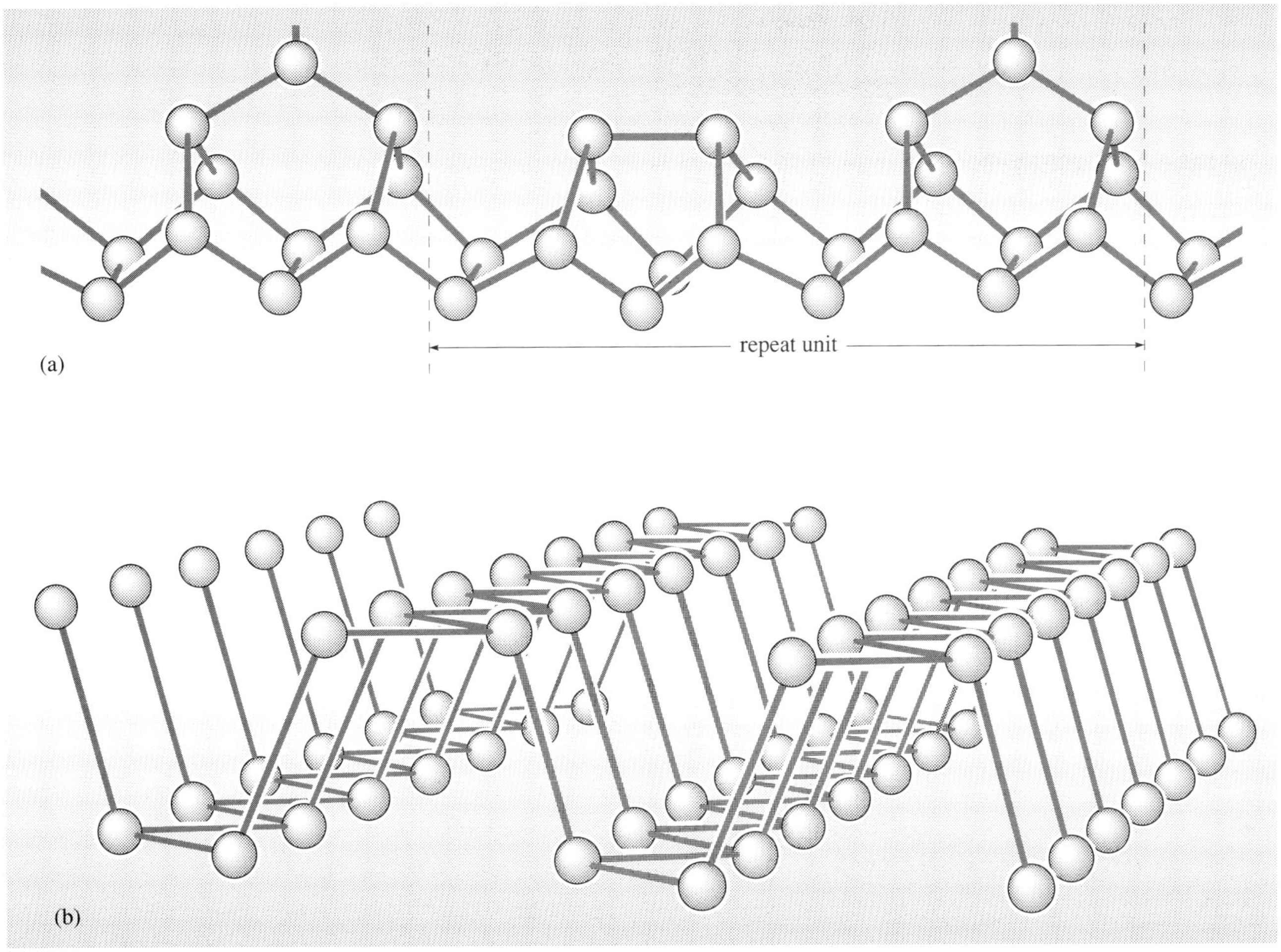

Figure 53 Structures of elemental polymeric phosphorus: (a) violet phosphorus has a complex double-layer structure with a repeat unit of 21 atoms; (b) black phosphorus has a layer structure, consisting of a puckered hexagonal net.

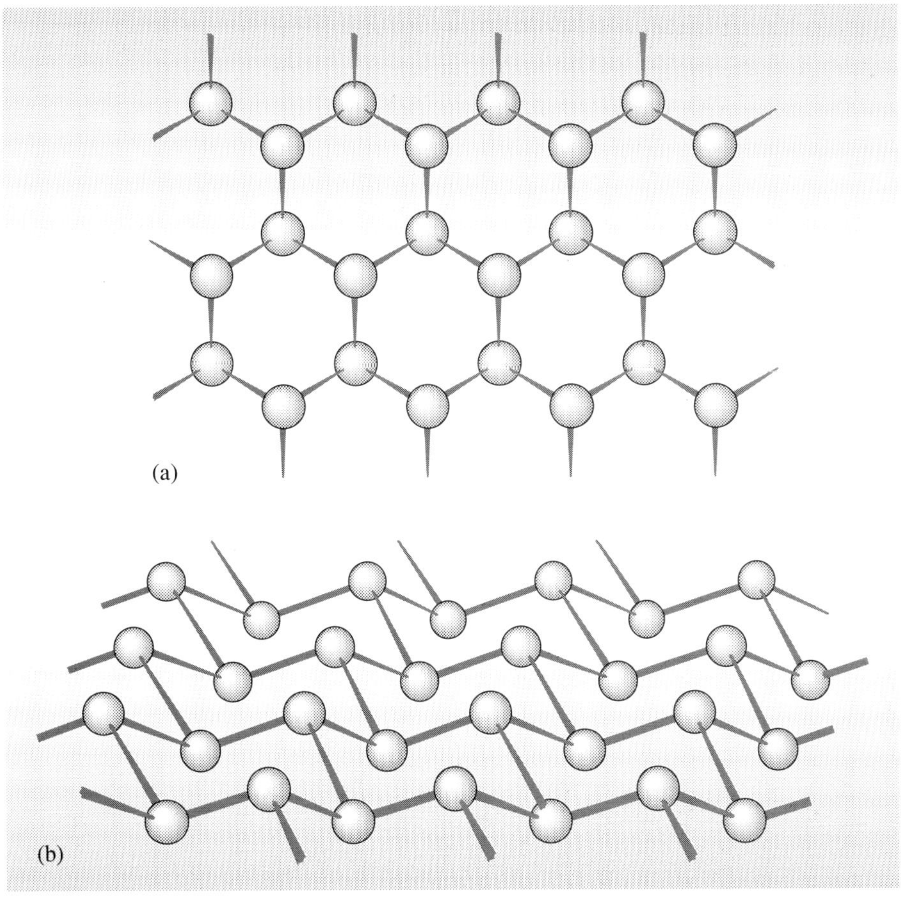

Figure 54 The isostructural α-forms of arsenic, antimony and bismuth. (Two perspectives are shown for clarity.)

Table 8 shows the gradation down the Group in the ratio of the distance between neighbours in the same layer and neighbours in adjacent layers. In close-packing, which is characteristic of metal structures, this ratio is unity.

Table 8 Interatomic distances in Group V elemental structures

Element		Distance from nearest neighbours (a)/pm	Distance from neighbours in next layer (b)/pm	Ratio (b/a)
black phosphorus		218	368	1.69
isostructural	α-arsenic	251	315	1.25
	α-antimony	291	336	1.15
	α-bismuth	310	347	1.12

5.2 NITROGEN

Nitrogen

'O' is for Oxygen
so gregarious
whereas I am
colourless
odourless
and tasteless
unattractive you might say
unreactive in every way
nitrogen: the night
to oxygen's day
I am 75%
of the air you breathe
so keep me clean.
For when I latch on
to fumes that cars exhaust
I am poison
Nitro-glycerin
that's me as well Dynamite
I can blow you all to hell
But I'm not without
a sense of humour
N_2O is the proof, nitrous oxide
Inhale some laughing gas
and see my funny side
N is my symbol
N for nebulous
necessary
and nondescript.

Roger McGough

Nitrogen is surprisingly versatile in its compounds: it can be found in oxidation numbers from −3 to +5.

■ What are the oxidation numbers of nitrogen in the nitride ion, N^{3-}, ammonia, NH_3, nitrogen fluoride, NF_3, and the nitrate ion, NO_3^-?

□ −3, −3, +3, and +5, respectively.

Like the other elements in the second row, it is able to form pπ bonds both to itself and to some of the other elements of the row, such as B, C and O. It is this property, together with a maximum coordination number of four, that makes its chemistry rather different from the rest of Group V.

Elemental nitrogen is fairly inert, with a formal triple bond, N≡N, and molar bond enthalpy of 945 kJ mol^{-1}; among the relatively few chemical reactions of molecular nitrogen at room temperature is the combination with lithium to form a red ionic compound, lithium nitride, Li_3N (recall Videocassette 2, Sequence 10). Many elements combine with nitrogen or ammonia on heating, to form nitrides of various types. The nitrides of the Group II metals are usually colourless (although that of magnesium is yellow), transparent and salt-like, and contain the nitride ion, N^{3-}. They are hydrolysed by water, to give the metal hydroxide and ammonia:

$$M_3N_2(s) + 6H_2O(l) = 3M(OH)_2(s) + 2NH_3(g) \qquad (53)$$

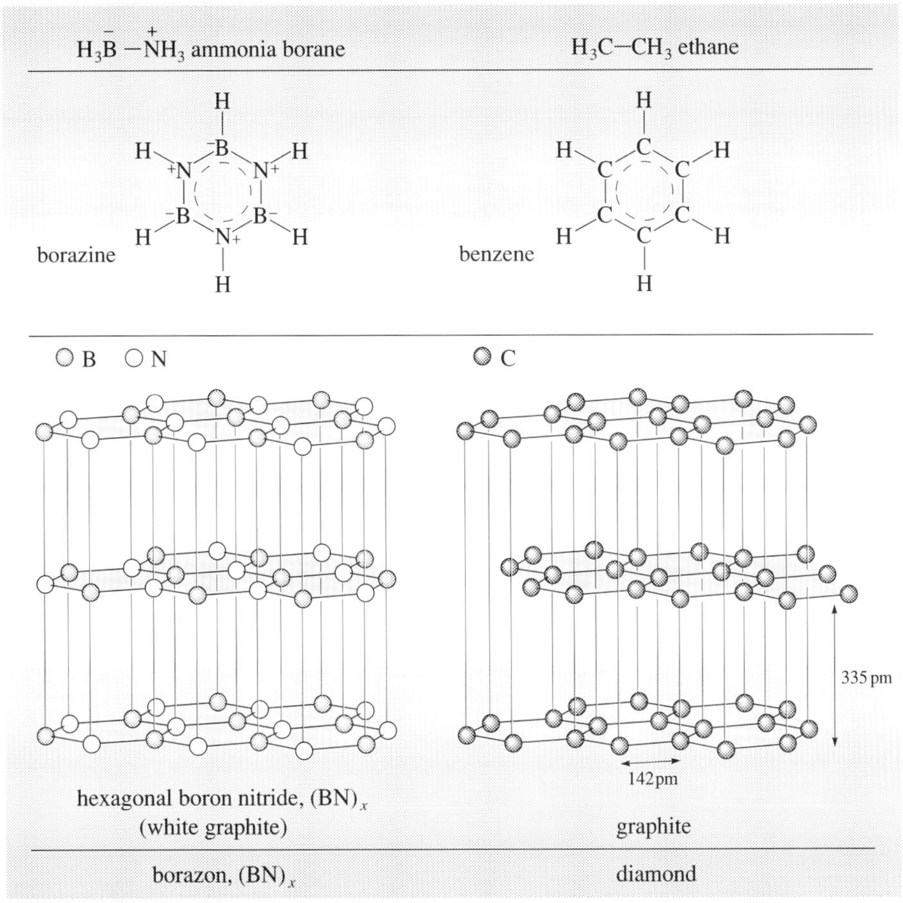

Figure 55 Boron–nitrogen compounds and their carbon analogues.

However, the nitrides of boron and aluminium are rather different: they are refractory materials with a macromolecular structure. The B—N grouping, with (3 + 5) valence electrons, is isoelectronic with C—C. Many boron–nitrogen compounds have been made that are analogues of the corresponding organic compounds: pairs of carbon atoms are replaced by B—N groups (Figure 55). Thus, $B_3N_3H_6$ is a structural analogue of benzene, although its chemical properties are rather different. Boron nitride, $(BN)_x$, has a hexagonal layer structure similar to graphite, the difference being that in $(BN)_x$ the layers lie directly over one another. Unlike graphite it is colourless and does not conduct electricity. This may be a factor of the different structure: because boron atoms lie over nitrogen atoms, the π electrons might be localized by $\overset{+}{N}—\overset{-}{B}$ interactions. If $(BN)_x$ is subjected to heat and pressure (1 800 °C and 85 000 atm), its structure changes to a cubic form analogous to diamond, known as borazon; this form is refractory and is used as an abrasive. It remains to be seen whether a form analogous to buckminsterfullerene can be made; it has been predicted theoretically that $B_{30}N_{30}$ should be stable. As we noted earlier for the silicon–oxygen and silicon–halogen bonds, the polarity of the bond lends extra strength to the $\overset{\delta+}{N}—\overset{\delta-}{B}$ bond compared with the C—C analogue, due to the attraction of opposite charges.

5.2.1 NITROGEN HYDRIDES

Ammonia, NH_3, is a colourless gas with a very strong characteristic smell (it is the gas given off by smelling salts); it also makes your eyes water. It reacts with hydrogen chloride to form a white smoke, which is composed of small particles of ammonium chloride: the laboratory test for ammonia is to use a glass rod dipped in concentrated hydrochloric acid and look for white fumes:

$$NH_3(g) + HCl(g) = NH_4Cl(s) \qquad (54)$$

Ammonia forms an alkaline solution in water, which turns red litmus blue:

$$NH_3(g) + H_2O(l) = NH_4^+(aq) + OH^-(aq) \quad (55)$$

As a weak base it is often used in household cleaners.

Ammonia is produced commercially by the Haber–Bosch process in vast quantities—more than any other compound—from the direct combination of the elements:

$$N_2(g) + 3H_2(g) = 2NH_3(g); \quad \Delta H_m^{\ominus} = -92\,\text{kJ mol}^{-1} \quad (56)$$

This reaction is very slow at room temperature, and the industrial plants operate at high temperature (400–450 °C) and high pressure (80–350 atm) in the presence of an iron catalyst. The ammonia is removed from the gas stream and can then be reacted with either nitric acid or sulphuric acid to make ammonium nitrate, NH_4NO_3, and ammonium sulphate, $(NH_4)_2SO_4$, respectively, which are largely used as fertilizers.

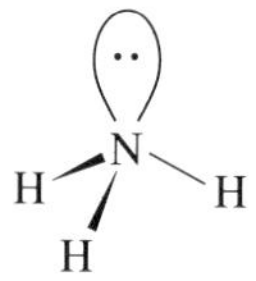

21

With nitrogen, we reach the first element of the second row to have a lone pair of electrons in its normal valency state; this is reflected in the structure of the ammonia molecule predicted by VSEPR theory and shown in Structure **21**.

Nitrogen is also among the most electronegative elements in the Periodic Table. We have described hydrogen-bonding as a mainly electrostatic attraction between a lone pair on an electronegative element and the hydrogen of a neighbouring atom. In Block 2 we saw how this type of bonding had a profound effect on the physical properties of ice, giving it a very open structure and consequently a low density; in Block 5 you saw that hydrogen-bonding accounts for a similar structure being adopted by solid ammonium fluoride. Hydrogen-bonding also affects the properties of ammonia: as a rule, we would expect the boiling temperatures of similar compounds to increase down a Group as the intermolecular forces increase. However, we find that the value for NH_3 is anomalous (NH_3, −33.4 °C, PH_3, −87.7 °C, AsH_3, −62.4 °C, SbH_3, −18.4 °C), being higher than we would expect from the trend for the rest of the Group V hydrides. This anomaly is explained by the extra intermolecular forces between the ammonia molecules as a result of hydrogen-bonding (see Block 5, Figure 18).

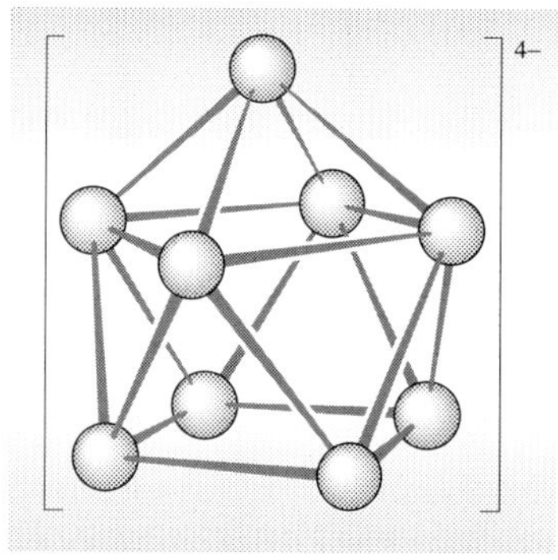

Figure 56 The C_{4v} structure of the Sn_9^{4-} anion, in which one atom is outside (or 'caps') one face of the square antiprism.

Ammonia forms a liquid at −78 °C, and can be used as a solvent; solutes include a number of metals. Dilute solutions of alkali metals in liquid ammonia are bright blue, paramagnetic and highly conducting. It has been found that the alkali metals ionize in the solution to give a cation solvated by ammonia molecules, M^+(am), and an electron, which is in a cavity of ammonia molecules with a radius of 300–340 pm, e^-(am). It is this solvated electron that has an absorption band giving rise to the deep blue colouration, and it is characteristic of the system whichever alkali metal is dissolved. At high concentrations the solutions become bronze-coloured and behave like molten metals. Solutions of alkali metals in ammonia can act as useful reducing agents, and are used in reactions forming unusual metal clusters such as Sn_9^{4-} (Figure 56) and Ge_9^{4-} among others.

Hydrazine, N_2H_4, is a fuming colourless liquid (m.t. 2 °C; b.t. 114 °C). In the gas phase the principal form is the gauche conformation (Structure **22**).

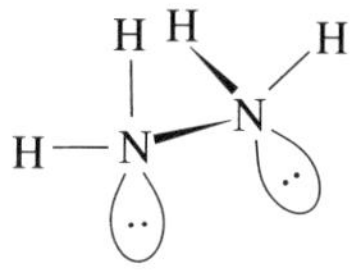

22

Hydrazine is produced by the **Raschig process**, in which ammonia is oxidized by sodium hypochlorite in the presence of gelatin (which suppresses side reactions)

$$2NH_3(aq) + OCl^-(aq) = N_2H_4(aq) + Cl^-(aq) + H_2O(l) \quad (57)$$

It has some industrial uses due to its reducing properties, but it—and its methyl derivatives, $H_2N{-}N(CH_3)_2$ and $H_2N{-}NH(CH_3)$—is mainly produced for use as a rocket fuel. Its utility for this purpose depends on its reducing ability. In the lunar landings, a 1 : 1 mixture of the methyl derivatives was mixed with dinitrogen tetroxide, N_2O_4, which acts as the oxidizing agent; this mixture ignites on contact. Equation 58 indicates the reaction for the dimethyl derivative:

$$H_2N{-}N(CH_3)_2(l) + 2N_2O_4(l) = 3N_2(g) + 4H_2O(g) + 2CO_2(g) \quad (58)$$

Other industrial uses include the silvering of mirrors by reducing solutions of silver salts, and the treatment of boiler water in electrical generating plants, to prevent corrosion of the boiler and pipework by dissolved oxygen in the water.

5.2.2 NITROGEN HALIDES

NF_3 is the most stable of the nitrogen halides; it does not react with water or dilute acid. It is prepared by the direct reaction of ammonia and fluorine:

$$4NH_3(g) + 3F_2(g) = 3NH_4F(s) + NF_3(g) \qquad (59)$$

The trichloride and bromide are very explosively unstable with respect to their constituent elements.

No pentahalides are known, and although the nitrogen(V) species NF_4^+ has been formed, we see again the inability of the second-row elements to exceed a coordination number of four in their compounds.

5.2.3 NITROGEN–OXYGEN COMPOUNDS

The nitrogen oxides, oxoions and oxoacids form a remarkable variation on the theme of σ- and π-bonding, which is elegantly described by molecular orbital theory. All are linear or flat, with π orbitals covering the whole molecule. We began their study with the simpler species (NO, NO^+, etc.) in Block 4, and continue here with the polyatomic species.

There are seven molecular oxides, and all are thermodynamically unstable with respect to the formation of N_2 and O_2: their structures (showing both the atom linkages and the most likely resonance forms) and physical properties are summarized in Table 9. You saw some of the nitrogen oxides in Videocassette 2, Sequence 10.

Table 9 The oxides of nitrogen

Formula	Name (trivial name in parentheses)	Linkage/structure				Physical properties
N_2O	dinitrogen monoxide (nitrous oxide)	N—N—O, linear, $\mathbf{C}_{\infty v}$				colourless gas (b.t. −89 °C)
NO	nitrogen monoxide (nitric oxide)	N—O, linear, $\mathbf{C}_{\infty v}$ N_2O_2, dimer	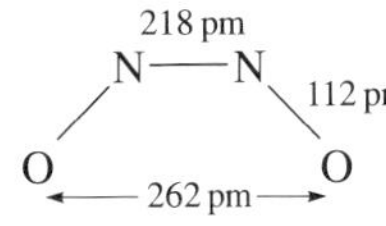	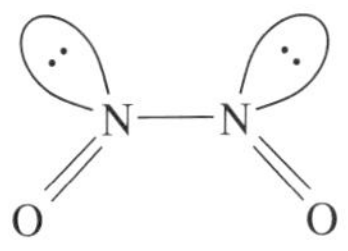		colourless paramagnetic gas (b.t. −152 °C)
N_2O_3	dinitrogen trioxide	planar, $\mathbf{C}_s$	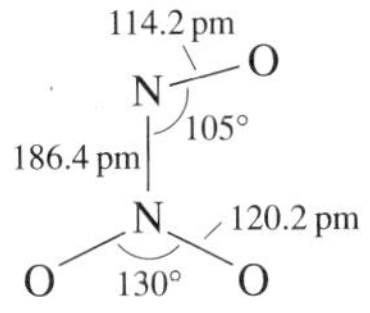	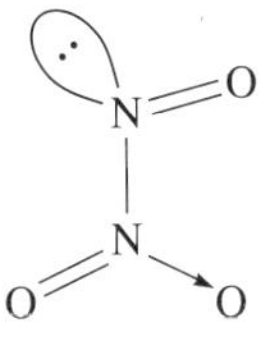	*or* 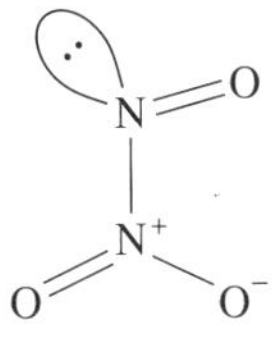	blue solid (m.t. −101 °C)
NO_2	nitrogen dioxide	bent, $\mathbf{C}_{2v}$	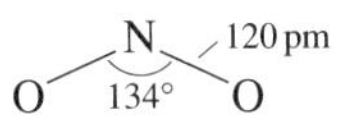	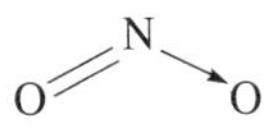	*or* 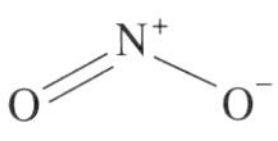	brown paramagnetic gas
N_2O_4	dinitrogen tetroxide	planar, $\mathbf{D}_{2h}$	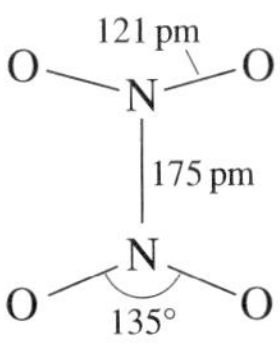	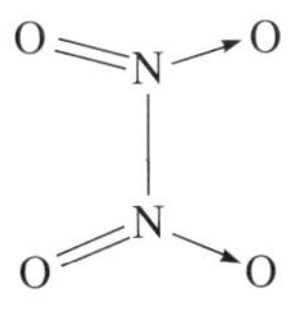	*or* 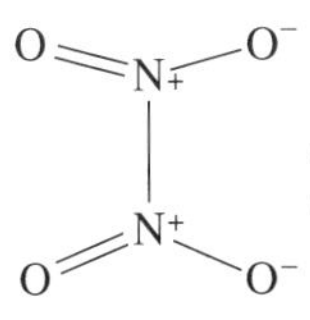	colourless liquid (m.t. −11 °C)
N_2O_5	dinitrogen pentoxide	ionic, $[NO_2]^+[NO_3]^-$				colourless solid (sublimes at 32 °C)
NO_3	nitrogen trioxide	trigonal planar, $\mathbf{C}_{3v}$				unstable paramagnetic radical

DINITROGEN MONOXIDE

Dinitrogen monoxide, N_2O, a non-toxic, odourless and tasteless gas, is a linear molecule with the unsymmetrical linkage N—N—O. It is made by the careful thermal decomposition of molten ammonium nitrate at 250 °C:

$$NH_4NO_3(l) = N_2O(g) + 2H_2O(g) \qquad (60)$$

On the whole it is rather unreactive, but has one rather useful and unusual reaction with molten alkali metal amides; this produces the alkali metal azide, containing the N_3^- ion. The reaction is used industrially for the preparation of the azides used in detonators.

$$NaNH_2(l) + N_2O(g) = NaN_3(s) + H_2O(g) \qquad (61)$$

Because it is almost non-polar, N_2O dissolves in fats; it finds application in the aeration of soft icecream and as the propellant in cans of whipped cream. It is best known as an anaesthetic, commonly called *laughing gas* because of its after-effects. It is the only gas apart from oxygen that will relight a glowing splint.

NITROGEN MONOXIDE

Nitrogen monoxide, NO, is one of the most reactive of the nitrogen oxides. It is an odd-electron molecule and is therefore paramagnetic (Block 4, Section 7.4); it has a bond order of $2\frac{1}{2}$. NO is a monomeric, colourless gas, which reacts immediately with atmospheric oxygen to give the characteristic brown colour of gaseous NO_2. When NO is cooled to a liquid, some N_2O_2 dimers are formed; as indicated in Structure **23**, these mainly have the *cis* structure. (The bond lengths come from the X-ray diffraction studies of the solid.)

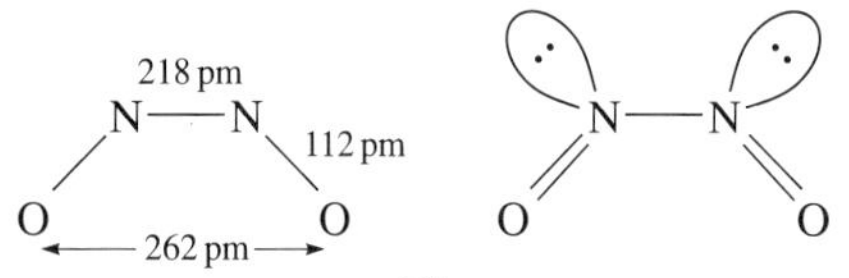

23

It is prepared in the laboratory by using a mild reducing agent on a nitrogen–oxygen compound in which the nitrogen has a higher oxidation number. Thus, Equation 62 shows the formation of nitrogen monoxide from nitrite ion, NO_2^-, and iodide ion in acid solution:

$$NO_2^-(aq) + I^-(aq) + 2H^+(aq) = NO(g) + \tfrac{1}{2}I_2(aq) + H_2O(l) \qquad (62)$$

Like CO, NO forms many coordination complexes with transition metals, when it is known as a **nitrosyl ligand**.

DINITROGEN TRIOXIDE

Dinitrogen trioxide, N_2O_3, can only be isolated at low temperatures (m.t. −101 °C) as a blue solid and deep blue liquid, when stoichiometric amounts of NO and NO_2 are combined. The molecule is planar (Structure **24**) with $\mathbf{C_s}$ symmetry. As the temperature is raised, the liquid becomes greenish as it disproportionates to NO and the brown NO_2 (see Plate 7b):

$$N_2O_3(l) = NO(g) + NO_2(g) \qquad (63)$$

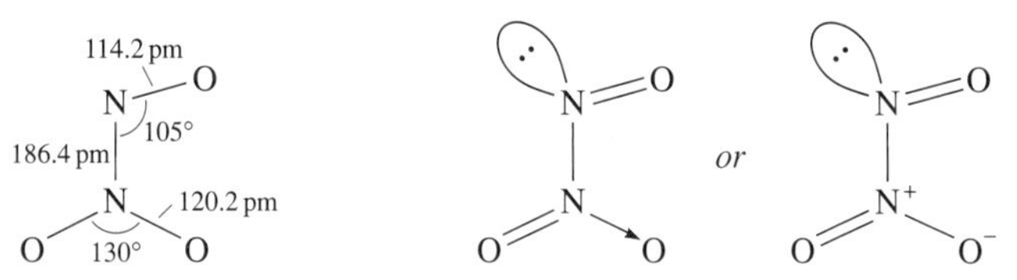

24

NITROGEN DIOXIDE AND DINITROGEN TETROXIDE

Nitrogen dioxide, NO_2, forms when NO reacts with O_2 or air; it is this reaction that is largely responsible for producing NO_2 in polluted air (the nitrogen oxides emitted by car exhausts are often given the general formula NO_x.)

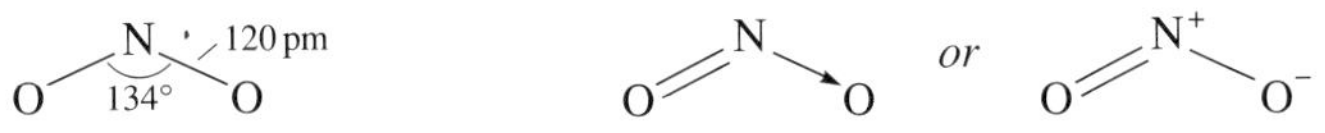

25

NO_2 is an odd-electron molecule (Structure **25**) and is therefore paramagnetic. The lone electron resides mainly on the nitrogen, which enables two molecules to interact to form a N—N bond. Thus, when NO_2 is cooled to form a liquid, it forms the dimeric N_2O_4 (Structure **26**).

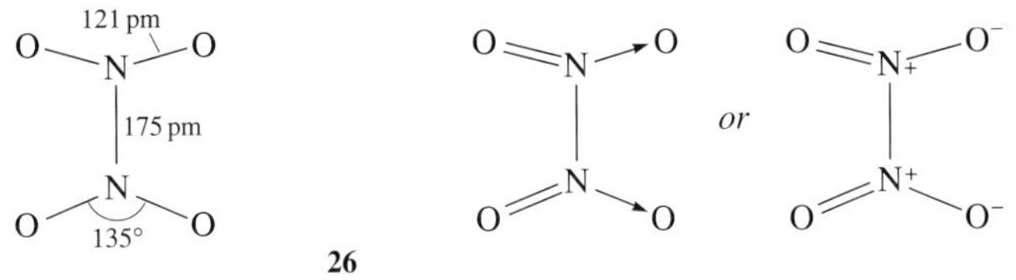

26

At the boiling temperature of N_2O_4 (21.5 °C) the mixture contains about 16% NO_2:

$$2NO_2(g) = N_2O_4(l) \quad (64)$$

The equilibrium between NO_2 and N_2O_4 was examined in Videocassette 1, Sequence 2 (see also Plate 7a).

NO_2/N_2O_4 are not only toxic but also corrosive because they react with water to form nitric acid:

$$N_2O_4(g) + H_2O(l) = HNO_3(aq) + HNO_2(aq) \quad (65)$$

$$3HNO_2(aq) = HNO_3(aq) + 2NO(g) + H_2O(l) \quad (66)$$

As you will read in Case Study 4, the production of NO_x by cars is still on the increase, and the nitric acid subsequently formed by them in the atmosphere is a continuing problem.

DINITROGEN PENTOXIDE

Dinitrogen pentoxide, N_2O_5, is the true anhydride of nitric acid, and it can be made by dehydrating concentrated nitric acid with P_4O_{10} at low temperature:

$$4HNO_3(aq) + P_4O_{10}(s) = 2N_2O_5(s) + 4HPO_3(aq) \quad (67)$$

It is a highly reactive, colourless solid, which has been shown to have an ionic structure, containing NO_2^+ and NO_3^- species.

NITROGEN TRIOXIDE

Nitrogen trioxide, NO_3, is a fleeting radical* species. It has been identified from its absorption spectrum, but has never been isolated as a pure compound.

Nitrogen forms many oxoacids; we consider here only the two most important ones.

* A radical is a fragment of a molecule, such as •OH, which has an unpaired electron; radicals tend to be very reactive.

NITROUS ACID

Nitrous acid, HNO_2, has not been isolated as a pure compound, although it is observed in equilibrium gaseous mixtures. It is a planar molecule, apparently preferring the *trans* structure shown in Structure **27**.

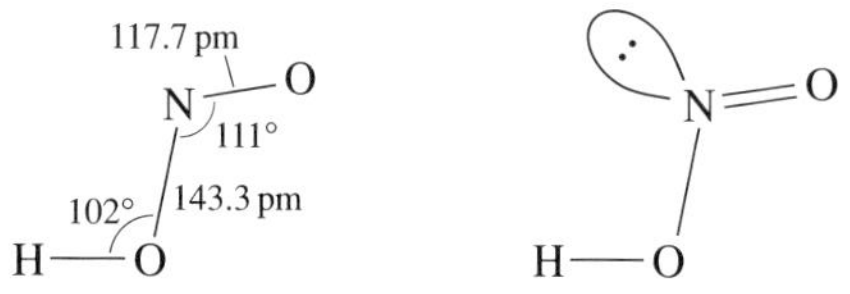

27

SAQ 22 *(Revising Block 4)* Assign the symmetry point group of the gaseous HNO_2 molecule.

In aqueous solution, HNO_2 is a fairly weak acid, dissociating to give the nitrite ion, NO_2^-. Sodium nitrite has been used for centuries for curing meat, particularly bacon and ham, although it is slightly toxic to humans.

NITRIC ACID

Nitric acid, HNO_3, is made on a huge industrial scale: approximately 20% of the ammonia produced each year is converted into HNO_3. This has many uses, but most (some 80%) is reacted with ammonia to produce ammonium nitrate, NH_4NO_3, for fertilizer; it is also used to make explosives such as TNT, nitroglycerine and nitrocellulose.

Nitric acid is produced by the **Ostwald process**: ammonia is oxidized in two stages over a catalyst made from platinum metals (Plate 8), first to NO and then to NO_2. The NO_2 is then dissolved in water to give a concentrated aqueous solution of the acid, and the NO produced in this step is recycled back into the earlier stages. The steps in the reaction can be summarized by the following equations:

$$NH_3(g) + \tfrac{5}{4}O_2(g) = NO(g) + \tfrac{3}{2}H_2O(g) \qquad (68)$$

$$NO(g) + \tfrac{1}{2}O_2(g) = NO_2(g) \qquad (69)$$

$$NO_2(g) + \tfrac{1}{3}H_2O(l) = \tfrac{2}{3}HNO_3(aq) + \tfrac{1}{3}NO(g) \qquad (70)$$

The anhydrous acid can be produced by distillation, and is a colourless pungent liquid. The nitric acid molecule is planar, as shown in Structure **28**.

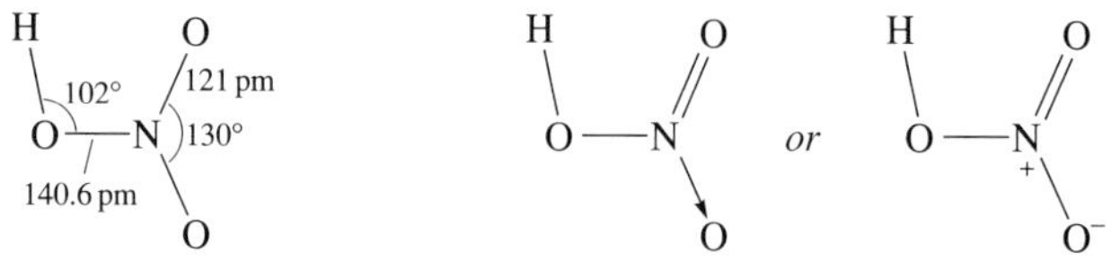

28

In dilute aqueous solution, nitric acid behaves as a typical strong acid, being extensively dissociated into H^+ and NO_3^- ions.

SAQ 23 *(Revising Block 4)* Predict the shape of the nitrate ion, NO_3^-, using VSEPR theory. What is its symmetry point group?

Ammonium nitrate, NH_4NO_3, mentioned above as an important fertilizer, is also rather unstable and has to be handled and packed with extreme care: it is thermally unstable at high temperatures, and its decomposition is catalysed by many inorganic and organic materials

$$2NH_4NO_3(l) = 2N_2(g) + O_2(g) + 4H_2O(g) \qquad >300\,^{\circ}C \qquad (71)$$

$$NH_4NO_3(l) = N_2O(g) + 2H_2O(g) \qquad 200\text{–}260\,^{\circ}C \qquad (72)$$

It is used extensively as an explosive by the mining and quarrying industries, where it is mixed with fuel oil.

Nitrogen/oxygen chemistry is explored further in the Home Experiments that follow.

STUDY COMMENT These experiments are assessed by a question in TMA 04.

5.3 HOME EXPERIMENTS: THE OXIDES OF NITROGEN

The oxidation numbers of nitrogen and oxygen provide a telling demonstration of the way in which stabilities are determined by a combination of thermodynamic and kinetic factors. Most of the Home Experiments that you will perform at the end of this Section are carried out in aqueous acid solution. They give you an opportunity to practise your predictive skills using redox potentials (Block 3, Appendix 1). You should have obtained some 20 vol. (6%) hydrogen peroxide and a gas cylinder for the Home Experiments because they are not supplied in the Home Kit.

5.3.1 COMPARISON OF THE THERMODYNAMICS OF REDOX REACTIONS IN AQUEOUS SOLUTION

The thermodynamic comparisons that tables of redox potentials allow are not restricted to metals and their aqueous ions. Redox potentials can be determined for any redox system in aqueous solution. Thus, in Table 10 we give values for a variety of systems, including halide/halogen reactions and more complicated redox systems with equations that include water and aqueous hydrogen ions. There are two important points to remember about a table of redox potentials such as this.

1 The strongest reducing agents are at the top of the Table; as we descend the Table, the thermodynamic strength of the reducing agent in the equations decreases, and that of the oxidizing agent increases.

2 If we have two redox systems, and the $E^\ominus$ value for the first is more positive than that for the second, then the oxidized state in the first is thermodynamically capable of oxidizing the reduced state in the second.

■ Assuming that reaction is not prevented by kinetic factors, would you expect a reaction between (a) $MnO_4^-(aq)$ and $I^-(aq)$; (b) $I_2(s)$ and $Fe^{2+}(aq)$? Give an equation for any reaction that occurs.

Table 10 Redox systems arranged in ascending order of $E^\ominus$ at 298.15 K

Redox system	$E^\ominus$/V
$K^+(aq) + e = K(s)$	−2.92
$Mg^{2+}(aq) + 2e = Mg(s)$	−2.36
$Al^{3+}(aq) + 3e = Al(s)$	−1.71
$Zn^{2+}(aq) + 2e = Zn(s)$	−0.76
$\frac{1}{2}I_2(s) + e = I^-(aq)$	0.53
$\frac{1}{2}O_2(g) + H^+(aq) + e = \frac{1}{2}H_2O_2(aq)$	0.68
$Fe^{3+}(aq) + e = Fe^{2+}(aq)$	0.77
$NO_3^-(aq) + 3H^+(aq) + 2e = HNO_2(aq) + H_2O(l)$	0.94
$HNO_2(aq) + H^+(aq) + e = NO(g) + H_2O(l)$	0.98
$\frac{1}{2}Br_2(aq) + e = Br^-(aq)$	1.10
$\frac{1}{2}Cl_2(g) + e = Cl^-(aq)$	1.36
$MnO_4^-(aq) + 8H^+(aq) + 5e = Mn^{2+}(aq) + 4H_2O(l)$	1.49
$\frac{1}{2}H_2O_2(aq) + H^+(aq) + e = H_2O(l)$	1.78
$\frac{1}{2}F_2(g) + e = F^-(aq)$	2.89

□ The $MnO_4^-|Mn^{2+}$ system lies below the $I_2|I^-$ system in Table 10: permanganate is a stronger oxidizing agent than iodine, and is thermodynamically capable of oxidizing iodide in acid solution. Note that the coefficient of e in the manganese reaction is 5. The first step in obtaining a balanced equation for the $MnO_4^-(aq)/I^-(aq)$ reaction is to multiply the iodine reaction by five. This is then subtracted from the $MnO_4^-(aq)/Mn^{2+}(aq)$ equation to eliminate e:

$$MnO_4^-(aq) + 5I^-(aq) + 8H^+(aq) = Mn^{2+}(aq) + I_2(s) + 4H_2O(l) \qquad (73)$$

In the second example, the $I_2|I^-$ system lies above, and has a less positive value of $E^{\ominus}$ than that for the $Fe^{3+}|Fe^{2+}$ system: iodine is a weaker oxidizing agent than $Fe^{3+}(aq)$. This means that the oxidation of $Fe^{2+}(aq)$ by iodine is thermodynamically unfavourable: the reaction cannot occur.

5.3.2 THE OXIDATION NUMBERS OF OXYGEN

In the presence of aqueous acid, oxygen can occur in oxidation number zero (oxygen gas), oxidation number −1 (aqueous hydrogen peroxide, H_2O_2) or oxidation number −2 (water). In hydrogen peroxide, therefore, oxygen is in an intermediate oxidation number. This means that hydrogen peroxide can behave either as a reducing agent or as an oxidizing agent. In the first case it is oxidized to oxygen; in the second, it is reduced to water. Abstracting the appropriate equations from Table 10, we have

$$\tfrac{1}{2}O_2(g) + H^+(aq) + e = \tfrac{1}{2}H_2O_2(aq); \qquad E^{\ominus} = 0.68\,V \qquad (74)$$

$$\tfrac{1}{2}H_2O_2(aq) + H^+(aq) + e = H_2O(l); \qquad E^{\ominus} = 1.78\,V \qquad (75)$$

The position of the two equations in Table 10 suggests that although hydrogen peroxide is a powerful oxidizing agent, it is not an especially strong reducing agent. In your Home Experiments you will see it act in both these roles.

Equations 74 and 75 also direct us to another interesting property of hydrogen peroxide. In Table 10 the equation in which hydrogen peroxide behaves as an oxidizing agent lies below that in which it acts as a reducing agent. Consequently, hydrogen peroxide can oxidize itself.

■ Write an equation for this reaction.

□ Subtracting Equation 74 from Equation 75, we get

$$H_2O_2(aq) = H_2O(l) + \tfrac{1}{2}O_2(g) \qquad (76)$$

Thus, hydrogen peroxide is thermodynamically unstable with respect to decomposition into water and oxygen gas. The solution is stable enough to be sold in pharmacists' shops, but as you will see in the Home Experiments, the decomposition can be speeded up by adding a catalyst such as manganese dioxide, MnO_2.

■ What other word can be used to describe this decomposition reaction?

□ It is a disproportionation. Oxygen in the intermediate oxidation number −1 is simultaneously oxidized and reduced to oxidation numbers 0 and −2.

5.3.3 THE OXIDATION NUMBERS OF NITROGEN

Compounds of nitrogen are known in which it is present in all nine integral oxidation numbers between −3 and +5 inclusive. Table 11 contains one example of a species representing each of the nitrogen oxidation numbers that can exist in contact with an aqueous acid solution. From a *thermodynamic* standpoint, the striking feature of nitrogen redox chemistry is the dominant stability of nitrogen gas (oxidation number zero). Thus, *all* of the binary oxides in Table 11 are thermodynamically unstable with respect to their constituent elements; for instance

$$2N_2O(g) = 2N_2(g) + O_2(g); \qquad \Delta G_m^{\ominus} = -208.4\,kJ\,mol^{-1} \qquad (77)$$

$$2NO(g) = N_2(g) + O_2(g); \qquad \Delta G_m^{\ominus} = -173.2\,kJ\,mol^{-1} \qquad (78)$$

$$2NO_2(g) = N_2(g) + 2O_2(g); \qquad \Delta G_m^{\ominus} = -102.6\,kJ\,mol^{-1} \qquad (79)$$

However, these reactions are not observed at room temperature; they occur at a detectable rate only on heating. Decomposition of dinitrogen monoxide to N_2 and O_2 begins above about 500 °C. Indeed, N_2O will relight a glowing splint, although not so readily as oxygen does. In the cases of NO and NO_2, formation of N_2 and O_2 is not observed below 1 000 °C; NO_2 first decomposes to NO and O_2 at a lower temperature (about 500 °C).

In aqueous acid, hydrazinium ion, hydroxylaminium ion and nitrous acid also have thermodynamically favourable decomposition reactions. Again, the most favoured route in a thermodynamic sense yields nitrogen as one of the products. All three decompositions are disproportionation reactions:

$$3N_2H_5^+(aq) + H^+(aq) = 4NH_4^+(aq) + N_2(g); \quad \Delta G_m^{\ominus} = -564.8\,\text{kJ mol}^{-1} \tag{80}$$

$$3NH_3OH^+(aq) = NH_4^+(aq) + N_2(g) + 3H_2O(l) + 2H^+(aq); \quad \Delta G_m^{\ominus} = -620.6\,\text{kJ mol}^{-1} \tag{81}$$

$$5HNO_2(aq) = 3NO_3^-(aq) + N_2(g) + H_2O(l) + 3H^+(aq); \quad \Delta G_m^{\ominus} = -293.1\,\text{kJ mol}^{-1} \tag{82}$$

Thus, of the species in Table 11, only $NH_4^+(aq)$, $N_2(g)$ and $NO_3^-(aq)$ are thermodynamically stable in contact with dilute acid. The fact that we can prepare and detect the other six in this medium indicates that, in nitrogen chemistry, conclusions based *solely* on thermodynamics are particularly unreliable. The most favoured thermodynamic pathway is often blocked by kinetic factors.

To take another example, when nitrous acid acts as an oxidizing agent, nitrogen gas is the most favoured product in a thermodynamic sense:

$$HNO_2(aq) + 3H^+(aq) + 3e = \tfrac{1}{2}N_2(g) + 2H_2O(l); \quad E^{\ominus} = 1.45\,\text{V} \tag{83}$$

but in practice, other products are often produced, the most common being nitrogen monoxide:

$$HNO_2(aq) + H^+(aq) + e = NO(g) + H_2O(l); \quad E^{\ominus} = 0.98\,\text{V} \tag{84}$$

It is this process which has been included in Table 10. There we also give the equation in which nitrous acid acts as a reducing agent and is oxidized to nitrate.

Like hydrogen peroxide, therefore, nitrous acid can act as either an oxidizing agent or a reducing agent. In the Home Experiments that follow, you will find out whether, like hydrogen peroxide, it disproportionates. If it does, you will then see if it disproportionates by the most thermodynamically favoured route, Equation 83, or by some other reaction.

Table 11 Nitrogen species that exist in contact with aqueous acid solutions

Species	Name	Oxidation number of nitrogen
$NH_4^+(aq)$	ammonium ion	−3
$N_2H_5^+(aq)$	hydrazinium ion	−2
$NH_3OH^+(aq)$	hydroxylaminium*	−1
$N_2(g)$	nitrogen gas	0
$N_2O(g)$	dinitrogen monoxide	+1
$NO(g)$	nitrogen monoxide	+2
$HNO_2(aq)$	nitrous acid	+3
$NO_2(g)$	nitrogen dioxide	+4
$NO_3^-(aq)$	nitrate ion	+5

* The hydroxylaminium ion is the conjugate acid of hydroxylamine, NH_2OH.

5.3.4 THE HOME EXPERIMENTS

Before you do Experiments 1–12, enter in Table 12 the products that you would expect from the redox potential data in Table 10. Note that such predictions are not those that would be made by relying *solely* on thermodynamics. This is because we have excluded from Table 10 some of the most thermodynamically favourable redox reactions involving nitrogen species. The reduction of aqueous nitrate to nitrogen gas, for example, is more favourable than the reduction to nitrous acid, whose equation is shown in the Table. We exclude it because it is rarely fast enough to be observed. Such exclusions are a way of combining the thermodynamic data with general observations on the kinetics of nitrogen redox reactions to produce a reasonably good predictive system. However, our general observations on the kinetics are not perfectly reliable, so you cannot assume that Table 10 will always enable you to predict the correct products.

In doing the aqueous experiments, you can assume that your predictions are unaffected by modest departures from the standard conditions to which the quoted values of $E^{\ominus}$ strictly apply, and by departures from a temperature of 298.15 K to which any heating gives rise. Note, however, that Table 10 refers to *acid solutions*. This is why we do not ask you to make predictions for Experiments 5, 6, 7 and 10, where reactions are carried out in the solid state, or in neutral or alkaline solution.

It will pay to take a little time to plan each experiment; decide first what the likely products are, and make sure that test reagents for these are to hand, since gas evolution may be short lived. When you are required to heat a reaction mixture in a test-tube, gently agitate the test-tube in the burner flame; use a small flame with the burner air holes nearly shut. These precautions will minimize the danger of liquid spurting from the test-tube. Take particular care when heating solid nitrates in experiments 6 and 7. Observe the tubes very carefully; some changes are easily missed. Make sure you can distinguish gas evolution on heating (a brisk stream of bubbles which continues on removal from the flame and is accelerated by shaking) from boiling. Finally, record all your observations carefully, even those that may surprise or disappoint you. This is particularly important, since your observations and conclusions will be tested in a TMA.

Summarize your experimental observations in Table 12, and your deductions from them. Make sure you know the colour and acidity of the possible products (see Section 5.2). Where relevant, write balanced equations for the reactions, noting the change in oxidation number of nitrogen.

The experiments are all done in test-tubes, and are not dangerous, but in several of them a gas is generated. *You should wear your safety spectacles and point the mouth of the test-tube away from yourself.* For Experiments 3, 5 and 8–11, the best heating method is to use water that has just boiled. If you heat it in the gas burner, shake the test-tube gently in order to achieve good mixing and to avoid hot spots.

The aqueous hydrogen peroxide is a 6% solution; as used in pharmacy, this may be labelled '20 vol.', referring to the volume of oxygen evolved when one volume of the solution is heated. The antiseptic and bleaching (which you can observe with a few hairs in a test-tube) actions are due to oxidation by the peroxide.

The reactions do not depend very critically on the amounts of material used. Liquids are measured roughly by depth in the test-tube. 'One measure' of a solid is the amount that can be balanced readily on the bent tip of a standard spatula. Test gases for acidity or basicity with wet litmus paper, red and blue, in the mouth of the test-tube (after making sure that the test-tube is clean). You may test for oxygen or N_2O with a glowing wooden splint, and for iodine with a slice of potato (or cornflour in water), which will turn blue.

Table 12 The Home Experiment redox reactions

Expt. No.	Reactants	Products		Conclusions
		Predicted by Table 10	Observed	
1	H_2O_2(aq)			
2	H_2O_2(aq) + Fe^{2+}(aq. acid)			
3	H_2O_2(aq) + I^-(aq. acid)			
4	H_2O_2(aq) + MnO_4^-(aq. acid)			
5	NH_4^+(aq) + NO_2^-(aq)	—		
6	NH_4^+(s) + NO_3^-(s)	—		
7	$Pb(NO_3)_2$(s)	—		
8	HNO_2(aq) + Fe^{2+}(aq. acid)			
9	HNO_2(aq. acid)			
10	$NaNO_3$ + Devarda's alloy* (aq. alkali)	—		
11	HNO_2(aq) + I^-(aq. acid)			
12	HNO_2(aq) + MnO_4^-(aq. acid)			

* Devarda's alloy is the best reagent for this reaction. It contains 50% Cu, 45% Al and 5% Fe.

HOME EXPERIMENT
OXIDATION NUMBERS OF NITROGEN AND OXYGEN

YOU NEED:

dropping pipette with teat
gas cylinder (not provided in the Kit)
gas burner nozzle
spatula
stirring rod
test-tubes
test-tube brush
test-tube holder
test-tube rack
wooden splints

ammonium chloride
ammonium iron(II) sulphate
ammonium nitrate
Devarda's alloy
hydrochloric acid (2M)
hydrogen peroxide (6% solution; not provided in the Kit)
lead nitrate
litmus paper, red and blue
manganese dioxide
potassium iodide
potassium permanganate
some form of starch (potato, for example; not provided in the Kit)
sodium hydroxide (2M)
sodium nitrate
sodium nitrite
sulphuric acid (2M)

All experiments are done in test-tubes.

1 Add a little manganese dioxide to 1 cm of aqueous hydrogen peroxide. (A speck of blood is an even better catalyst for this decomposition; cautious students may use a speck of fresh meat!)

2 Dissolve one measure of ammonium iron(II) sulphate in a minimum of water. Wet the end of the stirring rod in 2M sulphuric acid and dip the rod into your solution. Then add an equal volume of aqueous hydrogen peroxide and shake the mixture.

3 Dissolve a quarter measure of potassium iodide in 1 cm of 2M sulphuric acid. Add aqueous hydrogen peroxide dropwise until reaction is complete. Then pour off the liquid and heat the black solid at the bottom of the tube gently.

4 Dissolve one or two crystals (a minimal amount) of potassium permanganate in 1 cm of water and add 1 cm of 2M sulphuric acid. Put 1 cm of aqueous hydrogen peroxide in another test-tube, and add some of the permanganate solution dropwise, with shaking. (Save some permanganate solution for Home Experiment 12.) Note that acid permanganate solution is decolorized when reduced.

5 Dissolve two measures of sodium nitrite in 1 cm of water in a test-tube. In another test-tube dissolve one measure of ammonium chloride in 1 cm of water. Mix the two solutions and warm gently without boiling.

6 Put two measures of ammonium nitrate in a dry test-tube and heat, gently at first, and then more strongly. Note both gas and liquid products.

7 Heat three measures of lead nitrate gently, and then more strongly. The residual (glassy) solid is PbO.

8 Dissolve one measure of ammonium iron(II) sulphate and one measure of sodium nitrite in 1 cm of water. Wet the end of the stirring rod in 2M sulphuric acid and dip the rod into your mixture so that it absorbs the acid. Shake the mixture. Heat the solution gently, without boiling, and look at the gas in the test-tube against a white background, noting the colour immediately above the solution, and also at the top of the tube. (Compare this reaction with the one in Home Experiment 2.)

9 Dissolve one measure of sodium nitrite in 2 cm of 2M hydrochloric acid and observe what happens. Then heat the solution.

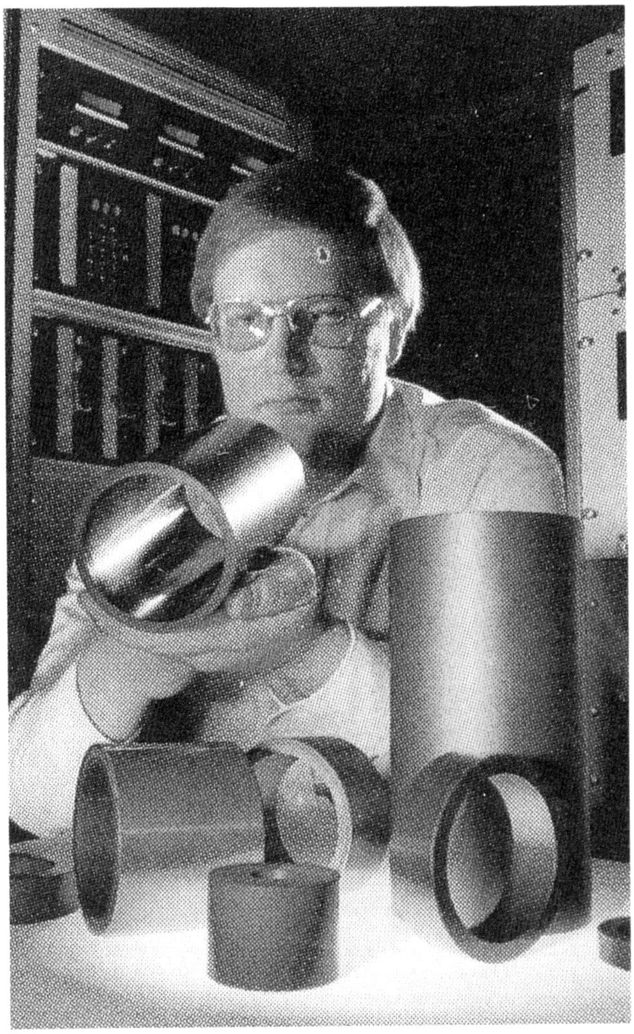

Plate 1 Silicon carbide, SiC, is used to manufacture high-strength, heat-resistant ceramics.

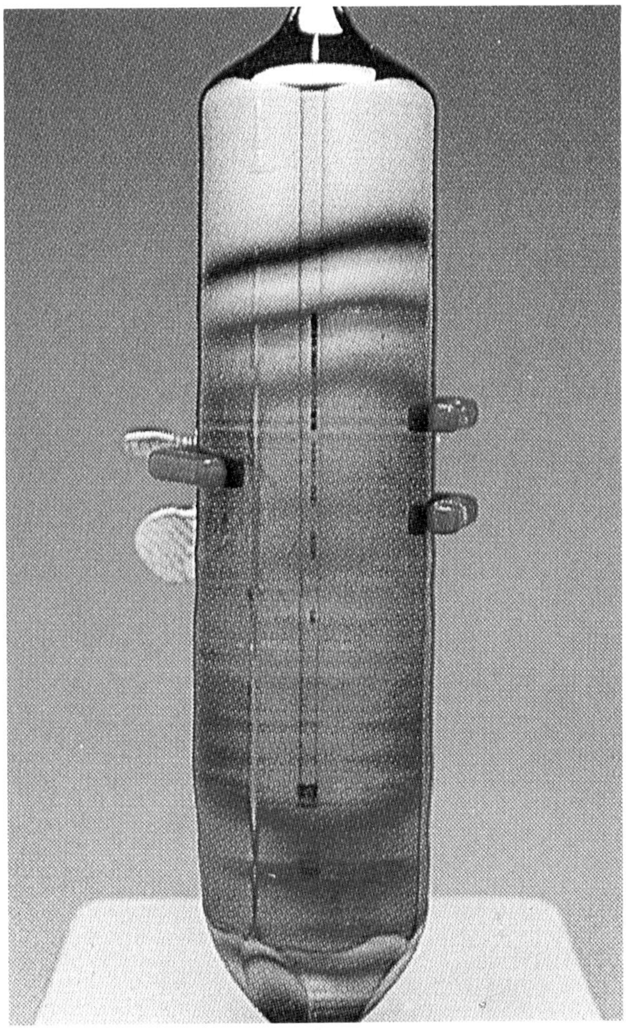

Plate 2 A rod of single-crystal silicon.

Plate 3 Natural crystals of quartz, SiO_2.

Plate 4 Beautifully formed crystals of aquamarine, heliodor, morganite and emerald.

Plate 5 A sample of zeolite, a silicate mineral. Zeolites are used for ion exchange, as molecular sieves and in catalysis.

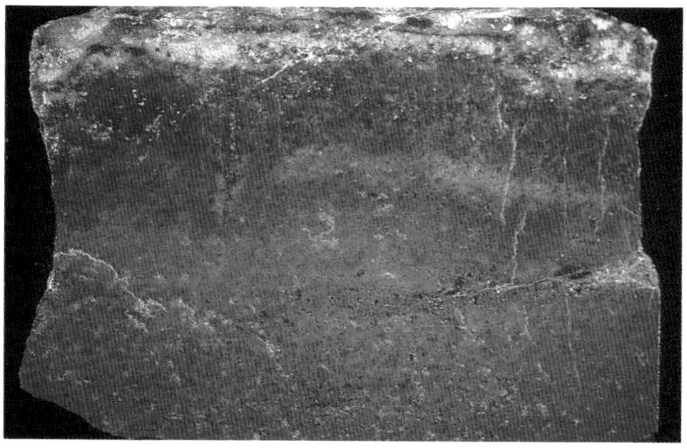

Plate 6 Polished slab of the silicate mineral lapis lazuli from Afghanistan, containing some brassy-coloured pyrite.

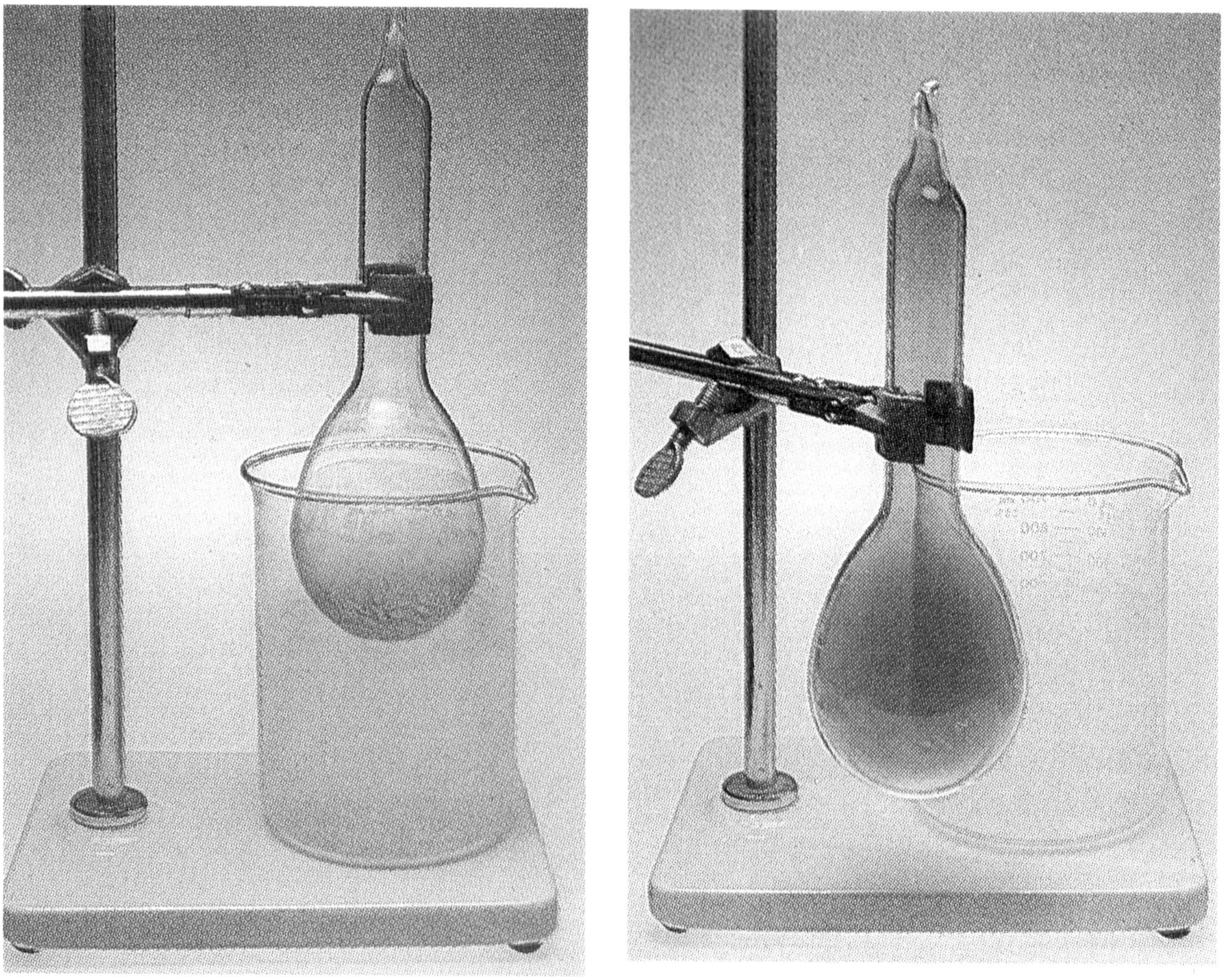

Plate 7 (a) The brown gas NO_2 in equilibrium with the colourless N_2O_4; (b) the sample contains a small amount of NO because, when it was frozen in liquid nitrogen, the blue colour of N_2O_3 appears.

Plate 8 The platinum metal catalyst for the oxidation of ammonia to nitrogen dioxide, the first stage of the Ostwald process for producing nitric acid from ammonia.

Plate 9 An example of eutrophic water in a lake near the Open University.

Plate 10 Iron pyrites, FeS_2, commonly known as fool's gold.

Plate 11 Orthorhombic sulphur crystals and crystalline calcite, found in a cavity in Sicilian limestone.

Plate 12 (a) When sulphur is heated above 150 °C the S_8 rings break open, forming chains, and the melt becomes viscous due to the tangling of the chains; (b) when the viscous melt is poured into water, plastic sulphur is formed, which gradually reverts to the orthorhombic form.

10 Dissolve one measure of sodium nitrate in 2M sodium hydroxide solution, and add not more than a quarter measure of Devarda's alloy. Using a test-tube holder, warm the mixture very gently; the reaction proceeds vigorously when hot.

11 As in Home Experiment 3, dissolve a quarter measure of potassium iodide in 1 cm of 2M sulphuric acid. In another test-tube, dissolve half a measure of sodium nitrite in 1 cm of water. Using a dropper, add a few drops of the sodium nitrite solution to the iodide solution and observe the reaction. When this is complete, pour off the liquid and heat some of the black solid at the bottom of the tube.

12 Dissolve one measure of sodium nitrite in 1 cm of water. Add the permanganate solution you saved from Home Experiment 4 dropwise, with shaking, to the nitrite solution.

5.4 SUMMARY OF SECTIONS 5.2 AND 5.3

1 Nitrogen, N_2, is an unreactive, gaseous, diatomic molecule with a triple bond. It has very high bond dissociation and ionization energies.

2 In combination with other elements, nitrogen is found in many oxidation numbers, from −3 to +5.

3 The Group II metal nitrides, M_3N_2, are salt-like in character, but BN and AlN have a macromolecular structure and are refractory in nature.

4 Ammonia, NH_3, is produced in great quantities by the Haber–Bosch process, largely to make fertilizers. When alkali metals dissolve in liquid ammonia, they form paramagnetic, conducting solutions due to the formation of solvated electrons.

5 Hydrazine, N_2H_4, and its methyl derivatives are used as rocket fuels.

6 The trihalides of nitrogen are all known, but tend to be rather unstable. The pentahalides are unknown.

7 Nitrogen forms seven molecular oxides; all the molecules are either linear or planar as a result of 2p–2p π-bonding. They are all thermodynamically unstable with respect to the elements.

8 Nitrogen forms two important oxoacids, nitrous acid, HNO_2, and nitric acid, HNO_3. Nitric acid is largely used to make the fertilizer NH_4NO_3.

9 In Table 10, redox systems are arranged in ascending order of the standard redox potential, $E^{\ominus}$, the strongest reducing agents being at the top of the Table.

10 For two redox systems, the one with a more positive value of $E^{\ominus}$ is thermodynamically capable of oxidizing the reduced state of the other.

11 Hydrogen peroxide can act either as an oxidizing agent of as a reducing agent.

12 The striking feature of nitrogen redox chemistry is the dominant stability of nitrogen gas.

13 The redox reactions of nitrogen in acid solution is often dictated by kinetic rather than thermodynamic factors.

5.5 PHOSPHORUS

Phosphorus was discovered by the German alchemist Hennig Brand of Hamburg in 1669. It was exhibited around the courts of Europe because of its ability to glow continuously, and without heat, in the dark. It was named after *phosphoros*, bringer of light, the name given by the ancient Greeks to the morning star. The secrecy surrounding Brand's method of preparation led others to try to prepare this remarkable new substance, and thus we find several men credited with rediscovering the element: all used urine as their source.

Since all living matter contains phosphate, PO_4^{3-}, in one form or another, phosphorus can in principle be obtained from any part of a plant or animal. Heating phosphate (most commonly as the calcium salt) with carbon — as a reducing agent — releases phosphorus as the volatile white phosphorus, P_4. Equation 85 gives this reaction:

$$2Ca_3(PO_4)_2(s) + 10C(s) = P_4(s) + 10CO(g) + 6CaO(s) \quad (85)$$

Whether the phosphate comes from urine or mineral ore, the reduction will take place if the temperature is high enough, and this reaction has been used as the basis for preparing the element for 300 years. Although urine was used for the first century of production, when most phosphorus was employed in medicine, this gave way to bones once it was found that bone is mainly the calcium phosphate called hydroxyapatite, $Ca_5(PO_4)_3(OH)$. Thus, bone ash served as the source of phosphorus for the newly established phosphorus industry. This grew up to supply the needs of the match industry from 1830–1910, during which period white phosphorus was used to initiate the chemical reaction of the match head: each match contained about 1 mg of P_4. The use of white phosphorus was eventually banned by international agreement because of the terrible industrial diseases it caused. Breathing air that was contaminated with phosphorus vapour could eventually lead to tooth decay that ate its way down to the jaw bone (known as 'phossy jaw'). White phosphorus was then replaced in matches by the less-volatile sulphide, P_4S_3.

Although bones provided the phosphate for the early industry, by 1875 mineral deposits of phosphates were being exploited, and these have been used ever since. Over two hundred phosphate minerals have been identified, but calcium phosphates form the major part of commercially exploitable deposits. One of the most important sources of phosphate is fluoroapatite, $Ca_5(PO_4)_3F$, which is found in several parts of the world, the two most important deposits being in Morocco (at Oulad-Abdoun) and in the USA (in Florida). Millions of tonnes are mined each year, mostly going to produce phosphate fertilizers (see Section 5.7). The other important use of the mined phosphate is in the production of the polyphosphate, $Na_5P_3O_{10}$, used in the detergent industry, as we shall see later in Section 5.6.5.

The present-day method for the industrial production of white phosphorus involves the reduction of mineral phosphate with coke and sand in an electric furnace:

$$2Ca_3(PO_4)_2(s) + 10C(s) + 6SiO_2(s) = P_4(g) + 6CaSiO_3(l) + 10CO(g) \quad (86)$$

The sand is added to the furnace along with the rock and coke, in order to effect the removal of the metal oxide as a fluid slag comprising mainly $CaSiO_3$.

White phosphorus has to be kept under water because it oxidizes spontaneously in air. As it oxidizes, it slowly warms up and inflames. It is soluble in organic solvents and reacts directly with many other elements to form (metal) phosphides, or oxides, sulphides, halides, etc., of phosphorus. White phosphorus is used for hardening and strengthening steel and bronze and for making phosphorus sulphide, used in the heads of non-safety matches (recall Block 5, Figure 59). Uses of phosphorus are summarized in Figure 57.

White phosphorus is heated by the tonne at 240 °C to turn it into red phosphorus, which is stable in air. This is mixed with powdered glass and a binder to make the strip on the sides of boxes of safety matches. Safety-match heads don't contain elemental phosphorus; they contain sulphur, the phosphorus sulphide, P_4S_3 (an important compound that we discuss more fully in Section 5.5.4), and oxidizing agents such as potassium chlorate, $KClO_3$.

The phosphate minerals are converted into phosphoric acid by industrial processes that will be discussed in Section 5.5.5. The industrial uses of phosphorus and its compounds are summarized in Figure 57. (Please make no attempt to memorize the information in this Figure: it is for interest only!)

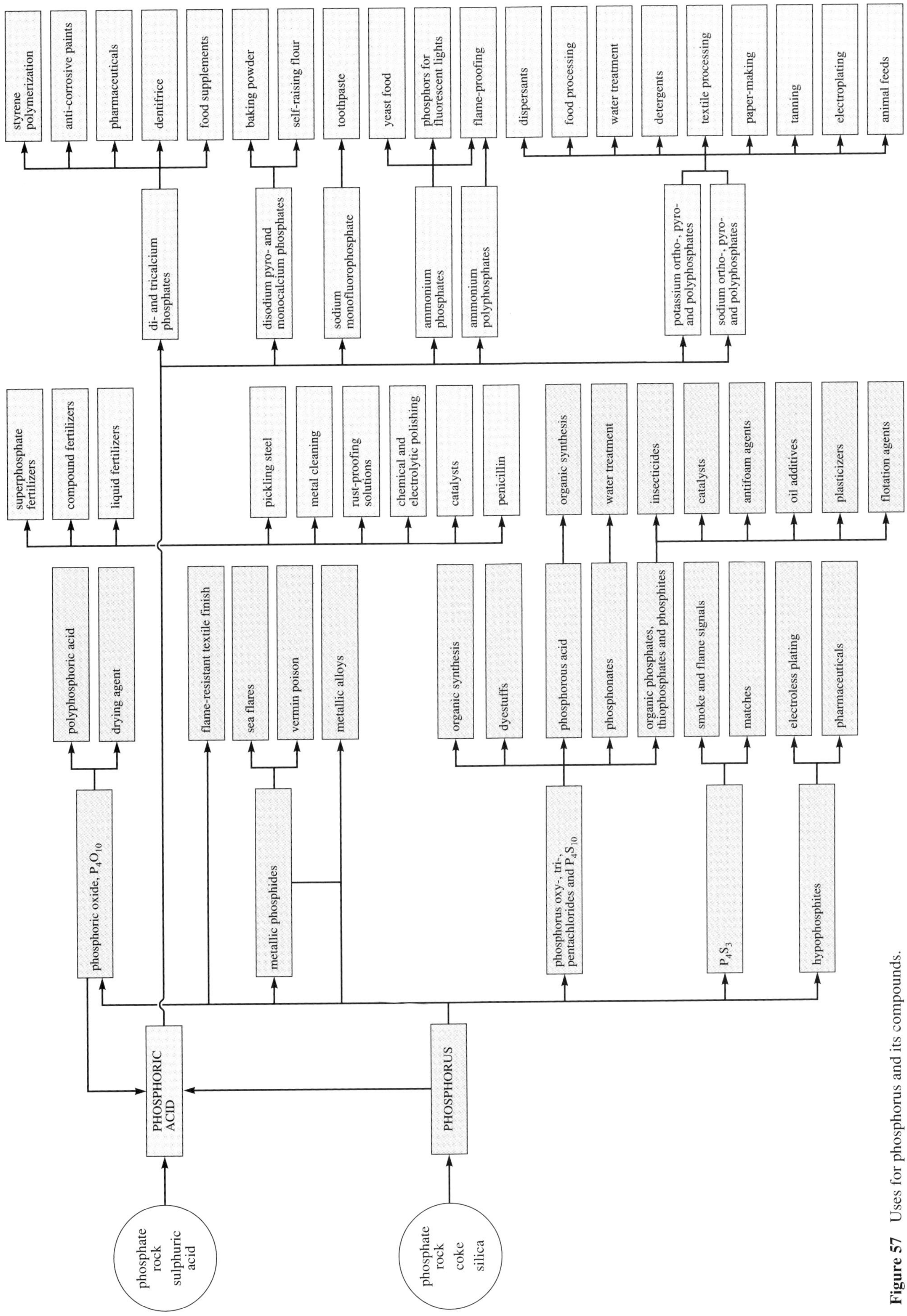

Figure 57 Uses for phosphorus and its compounds.

5.5.1 THE CHEMISTRY OF PHOSPHORUS

The electronic configuration of phosphorus is $[Ne]3s^23p^3$, and the common oxidation numbers of phosphorus with electronegative elements are +3 and +5. In oxidation number +3, phosphorus uses the three 3p electrons for bonding, leaving a lone pair of electrons. Because phosphorus is a third-row element, it can have a coordination number greater than four, but is also found with covalent bonding arrangements involving two, three, four, five or six atoms attached to phosphorus (Figure 58). In the trifluoride, PF_3, the oxidation number of phosphorus is +3.

■ Is the geometry of the PF_3 molecule depicted in Figure 58 what you would expect from VSEPR theory?

□ Yes. The three P—F single bonds and one lone pair constitute four repulsion axes, leading to a pyramidal structure. The effect of the lone pair will be to contract the tetrahedral angle slightly.

Phosphorus is found in oxidation number +5 in species such as $[PCl_4]^+$, PF_5, and $[PF_6]^-$.

■ How does the shape of $[PF_6]^-$ indicated in Figure 58 accord with that predicted from VSEPR theory?

□ There are twelve valence electrons (five from phosphorus, six from the fluorines and one negative charge). These are distributed to form six single bonds to fluorine in a regular octahedral shape. Note the expansion of the octet which is possible for third- and higher-row elements.

As we shall see in the following Sections, phosphorus also forms a number of **cage structures** based on expansion of the P_4 tetrahedron, containing three- or four-coordinate phosphorus. It has also been found to form ring or chain structures from linked tetrahedra with oxygen or nitrogen substituents on phosphorus.

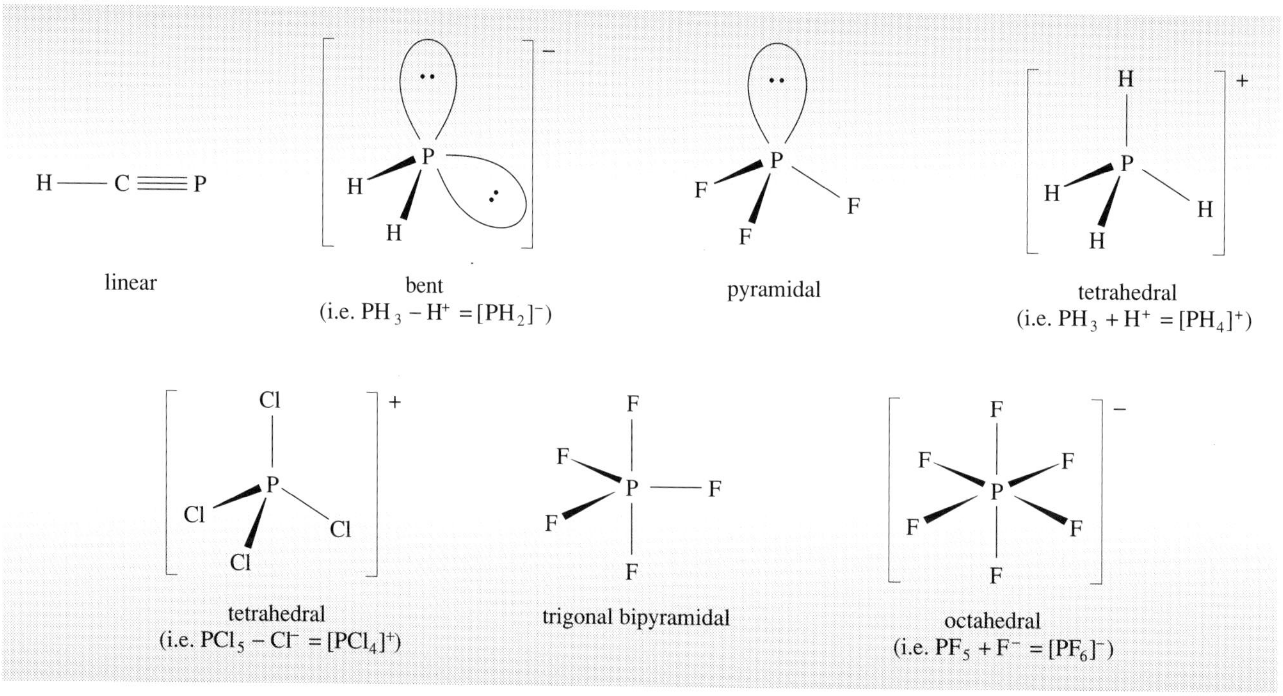

Figure 58 A selection of phosphorus-containing compounds, which illustrate the various coordination numbers that phosphorus can adopt in its compounds.

5.5.2 THE PHOSPHORUS HALIDES

Direct reaction of phosphorus and a halogen gives the trihalide (except PF_3) if phosphorus is kept in excess. For example, phosphorus and chlorine give PCl_3, which is a volatile, easily hydrolysed, fuming liquid:

$$P_4(s) + 6Cl_2(g) = 4PCl_3(l) \qquad (87)$$

Many other phosphorus compounds can be made from PCl_3. Industrially, it is important as a source of organophosphorus compounds for oil and fuel additives, plasticizers and insecticides.

PF_3 is a colourless, odourless gas, formed by the action of a metal fluoride such as CaF_2 on PCl_3:

$$3CaF_2(s) + 2PCl_3(l) = 2PF_3(g) + 3CaCl_2(s) \qquad (88)$$

It is poisonous because it coordinates to the haemoglobin in the blood in a similar fashion to CO. The trihalides are all rather volatile; even PI_3, which forms red crystals, has a melting temperature of 61 °C. All the trihalides are hydrolysed by water (although the hydrolysis of PF_3 is slow). The gaseous trihalides are pyramidal in shape.

All the PX_5 species can be formed; for example

$$P_4(s) + 10F_2(g) = 4PF_5(g) \qquad (89)$$

Phosphorus pentafluoride, PF_5, is a colourless gas.

■ Predict the shape of PF_5 according to VSEPR theory.

□ The molecules are trigonal bipyramidal in shape: five valence electrons on phosphorus and five from the fluorines form five electron pairs in P—F single bonds.

However, although phosphorus pentachloride, PCl_5, has the trigonal bipyramidal shape in the gaseous and liquid states, at room temperature it forms off-white crystals containing the ionic species $[PCl_4]^+[PCl_6]^-$ (Figure 58; $[PCl_6]^-$ has the same structure as $[PF_6]^-$). Phosphorus pentabromide forms reddish-yellow crystals, containing the ions $[PBr_4]^+Br^-$, but dissociates in the gas phase to PBr_3 and Br_2; the pentaiodide is thought to be similar.

The diphosphorus tetrahalides, P_2X_4, are also all known, as are many mixed halide compounds.

5.5.3 THE PHOSPHORUS HYDRIDES (PHOSPHANES)

The phosphides of electropositive metals such as calcium phosphide, Ca_3P_2, are easily hydrolysed to give phosphine, PH_3. Phosphine is a very reactive and poisonous gas; it has been described as smelling of both garlic and rotten fish! It tends to inflame spontaneously in air owing to the presence of small amounts of P_2H_4 and P_4 impurities. PH_3 is much less basic than NH_3, and although **phosphonium salts**, $[PH_4]^+X^-$, are known, only PH_4I is stable at room temperature; the other phosphonium halides dissociate to PH_3 and HX.

Further products of the hydrolysis of calcium phosphide are P_2H_4, P_3H_5 and other phosphanes, all of which contain P—P bonds. These are very sensitive to air and also are very unstable, tending to decompose to PH_3 and phosphorus-rich polymers. The most stable phosphane known is P_7H_3, which has the cage structure shown in Figure 59. Its stability is due to the wide separation of the hydrogen atoms, so that PH_3 is not lost readily. Replacement of hydrogen in phosphanes by organic groups such as methyl, generally makes them much more thermally stable. Numerous organophosphorus compounds are known.

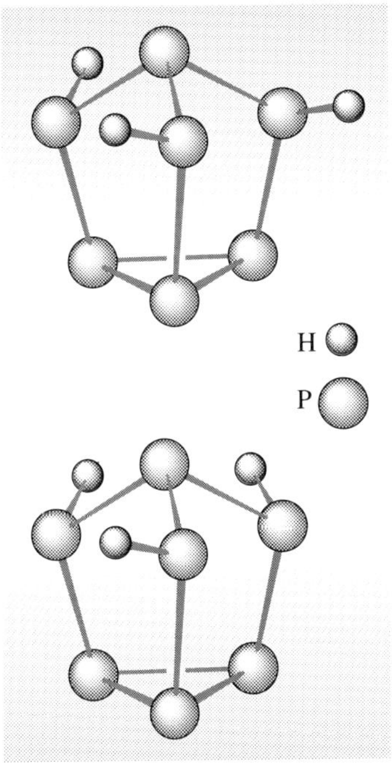

Figure 59 Two isomers of P_7H_3, which arise from different orientations of hydrogen atoms and lone pairs of electrons on the three phosphorus atoms in the middle of the cage.

5.5.4 THE OXIDES AND SULPHIDES OF PHOSPHORUS

Phosphorus reacts readily with oxygen and sulphur. The oxidation of P_4 by air or oxygen can give a ghostly white phosphorescence, which was formerly used in theatrical effects, but which is now banned because of fire and toxicity hazards. Controlled reaction with oxygen gives P_4O_6, a volatile molecular compound with a cage structure in which oxygen atoms have inserted into each P—P bond of P_4 (Figure 60a). This is commonly known as phosphor*ous* oxide (note the different spelling from the element, which is phosphor*us*). Further oxidation adds terminal oxygen atoms stepwise, giving P_4O_{10}, phosphoric oxide, as the final oxidation product. P_4O_{10} has the cage structure shown in Figure 60b. Phosphorous oxide melts at 24 °C, and phosphoric oxide sublimes fairly readily, a volatility we might expect for molecular solids containing discrete molecules of low molecular mass.

(You may also come across the names phosphorus trioxide and pentoxide, respectively, for P_4O_6 and P_4O_{10}. This stems from their empirical formulae, P_2O_3 and P_2O_5, which were known long before their structures were determined.)

■ What are the oxidation numbers of phosphorus in P_4O_6 and P_4O_{10}?

□ The oxidation numbers are, respectively, +3 (P^{III}) and +5 (P^{V}).

■ What are the coordination numbers of phosphorus in the two compounds?

□ P_4O_6 and P_4O_{10} have coordination numbers of three and four, respectively.

In the chemistry of the elements of the third row of the Periodic Table, we have seen many single-bonded structures, both in the elemental forms and in the oxides, which contrast with the multiply bonded structures formed by the second-row elements: for example, SiO_2 may be contrasted with CO_2 and P_4O_6 with N_2O_3. But with P_4O_{10} there is a significant change: each phosphorus has a terminal oxygen, and the terminal phosphorus–oxygen bond length is much shorter (143 pm) than the bridging phosphorus–oxygen bond (160 pm; see Figure 60b); hence the terminal phosphorus–oxygen groups may be thought of as double bonded. This contrast is also found in the phosphates (containing P^{V}) and the phosphites (containing P^{III}). In sulphur chemistry, we shall find similar behaviour.

P_4O_6 and P_4O_{10} are acid anhydrides of their respective oxoacids: when P_4O_6 reacts with water, it forms phosphorous acid (Equation 90), whereas P_4O_{10} forms phosphoric acid (Equation 91):

$$P_4O_6(s) + 6H_2O(l) = 4H_3PO_3(aq) \tag{90}$$

$$P_4O_{10}(s) + 6H_2O(l) = 4H_3PO_4(aq) \tag{91}$$

Phosphoric oxide finds ready application in the laboratory, for it is the most powerful **drying agent** known under ordinary conditions. The reaction with water is vigorous or violent, depending on the quantities involved. In fact, the 'craving' of phosphoric

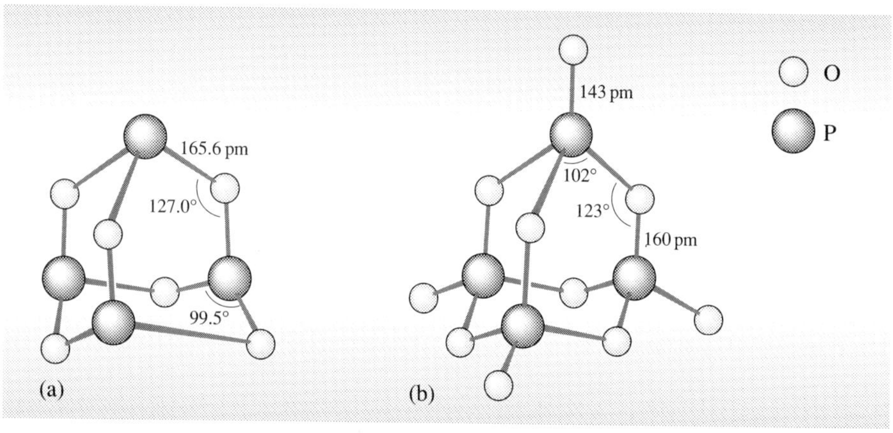

Figure 60 The structures of (a) phosphorous oxide, P_4O_6, and (b) phosphoric oxide, P_4O_{10}.

oxide for water is so great that it can be used as a **dehydrating agent**; that is, it will extract the elements of water from some compounds. It will even dehydrate compounds that are themselves good dehydrating agents, such as sulphuric acid, H_2SO_4, which is dehydrated to its acid anhydride, SO_3. The absorbed moisture first makes the phosphoric oxide sticky, and then forms acid solutions. An example of the use of P_4O_{10} as a dehydrating agent is the formation of a nitrile from an amide:

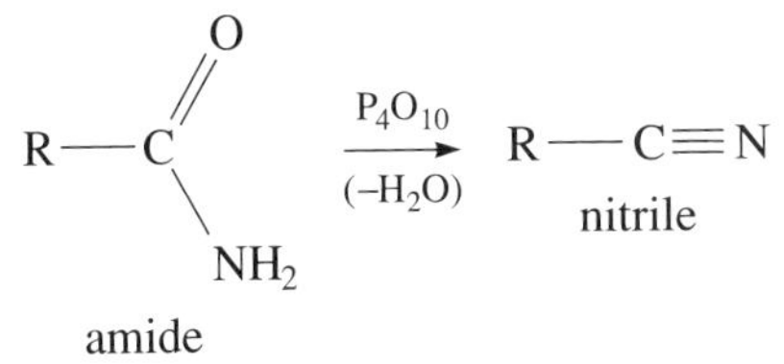

$$\text{amide} \qquad \xrightarrow[\text{(−H}_2\text{O)}]{P_4O_{10}} \qquad R\text{—}C\equiv N \qquad\qquad (92)$$

Reaction of phosphorus with a limited supply of sulphur yields P_4S_3, mentioned in Section 5.5 as a component of matches (see also Block 5, Figure 59). It has the same cage structure as P_7H_3, with each PH unit replaced by S as shown in Figure 61. The compounds P_7H_3 and P_4S_3 are isoelectronic. Further reaction with sulphur gives phosphorus sulphides with fewer P—P bonds and more P—S—P groups: the ultimate reaction product is a phosphorus(V) sulphide, P_4S_{10}, which has the same structure as P_4O_{10} (Figure 60b); it contains P—S—P bonds and a terminal P=S bond on each phosphorus atom.

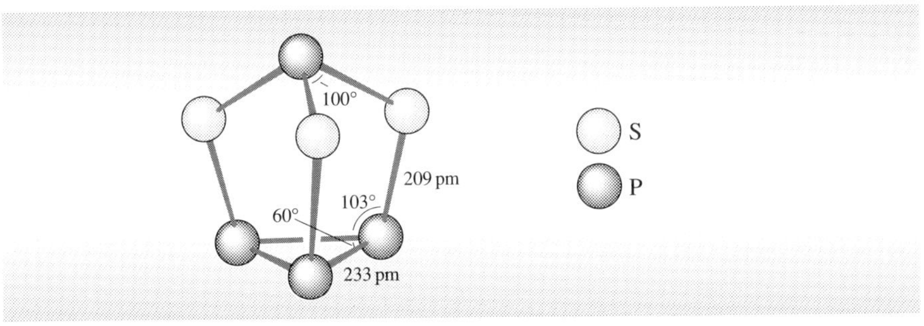

Figure 61 The structure of P_4S_3.

5.5.5 PHOSPHORIC ACID

Phosphoric acid is manufactured from phosphate minerals by two main industrial processes. The most important process (by an order of magnitude) is known as the **wet acid process**. It produces phosphoric acid, H_3PO_4, from the reaction of dilute sulphuric acid with a mineral phosphate (which we have regarded as being essentially calcium phosphate in Equation 93):

$$Ca_3(PO_4)_2(s) + 3H_2SO_4(aq) = 3CaSO_4(s) + 2H_3PO_4(aq) \qquad (93)$$

This produces a less pure product than the phosphoric acid produced by the thermal process described below, but is only about one-third the price. The gypsum (calcium sulphate) is filtered off with other solid impurities and the dilute phosphoric acid obtained is concentrated by evaporation to a rather impure dark brownish liquid, which is suitable for making phosphate fertilizers without further purification (see Section 5.7).

The **thermal process** forms an alternative route to phosphoric acid. It involves production of elemental phosphorus as an intermediate, and may be used when food-quality chemicals are required. The elemental phosphorus can be used for some compounds that cannot be made from phosphoric acid, such as the halides and sulphides. The first step in the process is Equation 86, the reduction of phosphate by carbon, in the presence of silica, which forms a removable silicate slag with the calcium. The reduction is carried out at $1\,600\,°C$ in an electric furnace. Phosphorus is removed as a vapour and condensed under water.

$$2Ca_3(PO_4)_2(s) + 10C(s) + 6SiO_2(s) = P_4(g) + 6CaSiO_3(l) + 10CO(g) \qquad (86)$$

Phosphoric acid can then be made by the oxidation of phosphorus and subsequent treatment of the phosphoric oxide, P_4O_{10}, with water:

$$P_4(s) + 5O_2(g) = P_4O_{10}(s) \tag{94}$$

$$P_4O_{10}(s) + 6H_2O(l) = H_3PO_4(l) \tag{95}$$

Pure phosphoric acid forms low-melting crystals (m.t. 42 °C); the structure of the H_3PO_4 molecule is shown in Figure 62.

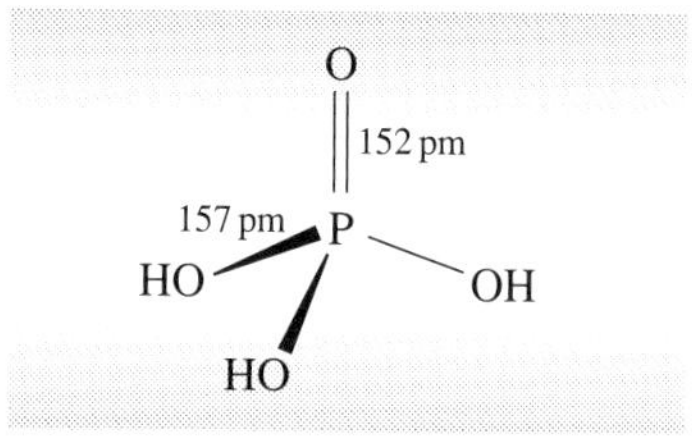

Figure 62 The structure of phosphoric acid.

Commercial phosphoric acid is 85% phosphoric acid in water; this forms a syrup, because the acid molecules are hydrogen-bonded to water molecules.

The salts and esters (alcohol derivatives) of phosphoric acid are important industrially and find a wide number of uses. In particular, the esters are used as insecticides, oil additives and in the recycling of uranium.

5.5.6 COMPOUNDS WITH MULTIPLE BONDS BETWEEN PHOSPHORUS ATOMS AND FROM PHOSPHORUS TO CARBON

We mentioned earlier (Section 5.1) that the P_2 molecule, with a phosphorus–phosphorus triple bond, is unstable except at high temperatures and that P_4 and other forms of phosphorus with P—P single bonds are more stable at ordinary temperatures. This behaviour is in contrast to that of nitrogen, which is stable only as N_2 molecules. The difference between the two elements is a consequence of 2p–2p π-bonding being more effective than 3p–3p π-bonding, and of the N—N σ bond being weaker than the P—P σ bond. The bond enthalpy terms are listed in Table 13. Inspection of those data shows that

Table 13 Molar bond enthalpy terms for some nitrogen and phosphorus bonds

Bond	B/kJ mol^{-1}
N≡N	945
N—N	158
P≡P	485
P—P	198

$$B(\text{N}{\equiv}\text{N}) > 3B(\text{N—N}), \text{ but}$$

$$B(\text{P}{\equiv}\text{P}) < 3B(\text{P—P})$$

Early attempts to make compounds containing phosphorus–phosphorus double bonds that were stable under normal conditions, all failed: only single-bonded compounds were formed. For instance, one reaction that was tried is shown in Equation 96 below. The object was the reduction of a methylphosphorus dichloride with sodium:

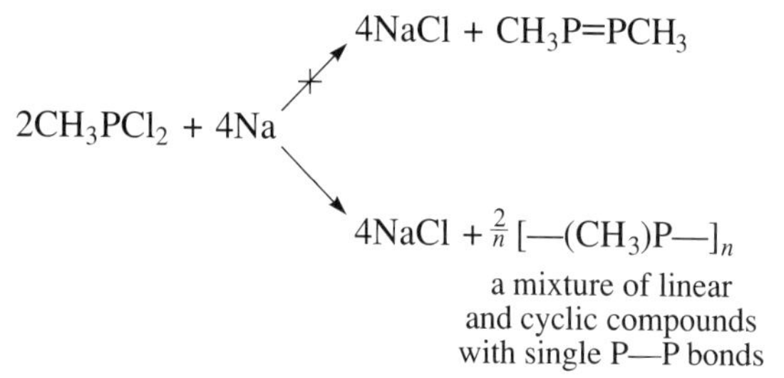

(96)

The compound $CH_3P{=}PCH_3$ probably forms as an intermediate, but is unstable with respect to polymerization.

■ From your knowledge of silicon chemistry, how would you try to stabilize a P=P bond?

□ The bulky mesityl group stabilized the Si=Si bond, so it would be reasonable to try the same strategy here.

In the late 1970s it was discovered that the reaction scheme in Equation 96 would produce the double-bonded product if the methyl group was changed for a much bulkier organic group, such as a tri-tertiarybutylphenyl group, as shown in Figure 63.

■ Why should a bulky substituent prevent the formation of P—P bonded polymers?

□ As we saw earlier in the formation of Si=Si double bonds, large organic groups introduce kinetic barriers, and prevent two molecules coming close enough together to react and form a polymer.

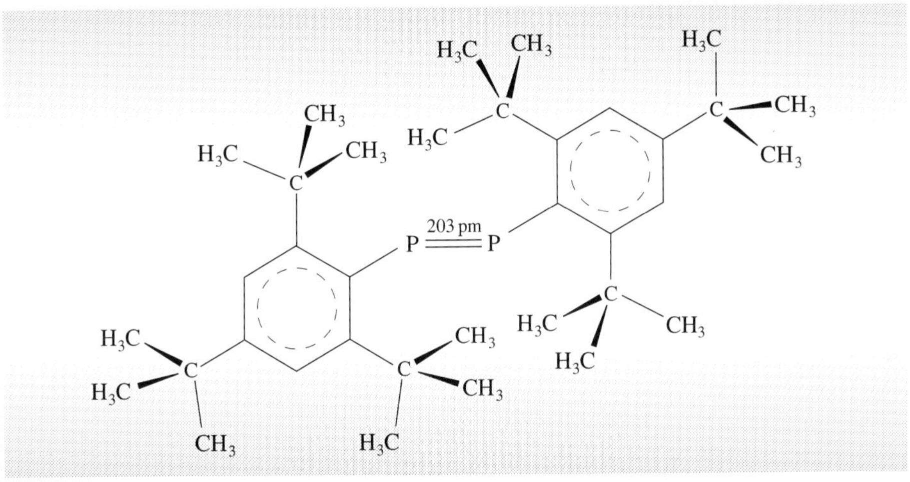

Figure 63 Structure of the $(2,4,6\text{-}(Bu^t)_3C_6H_2)_2P_2$ molecule.

The P—P bond length in the compound shown is 203 pm, much shorter than a typical P—P bond length of 220 pm. However, the bulky groups provide no protection from attack by small molecules, and so, for instance, the compounds are air-sensitive because of reaction with oxygen.

Bulky substituent groups have also been used to stabilize compounds containing phosphorus–carbon double or triple bonds. The bonding in this case is 2p–3p, which we would expect to be stronger than 3p–3p. Nevertheless, polymerization still tends to occur. Figure 58 (p. 68) shows the linear H—C≡P molecule, which is stable under low pressure in the gas phase, and has been detected in interstellar space, but it polymerizes rapidly when condensed. The tertiarybutyl analogue of this compound, $(CH_3)_3CC{\equiv}P$, is stable to polymerization at room temperature, although it is more chemically reactive than the corresponding alkyne, $(CH_3)_3CC{\equiv}CH$.

5.5.7 PHOSPHORUS–NITROGEN COMPOUNDS: THE POLYPHOSPHAZENES

■ The atom combination PN is isoelectronic with which Si species?

□ SiO.

We might expect, therefore, to find phosphorus/nitrogen polymers, $(R_2PN)_n$, analogous to the silicones, $(R_2SiO)_n$, we looked at in Section 4.3.4. These polymers are indeed known and are called **polyphosphazenes**; in recent years they have found various practical applications.

The parent compound of the phosphazenes is the cyclic compound, $Cl_6P_3N_3$, shown in Figure 64. The ring is almost planar, and although we have drawn it with alternating single and double bonds around the ring, this is just a formalism, and the experimental results show all the P—N bonds to be equal and rather shorter than expected for a P—N single bond. There is much debate about the bonding in these compounds because it has also been found to be rather different from the delocalized p_π–p_π systems such as is found in benzene, and while some authors favour contributions from the 3d orbitals on phosphorus, many others disagree.

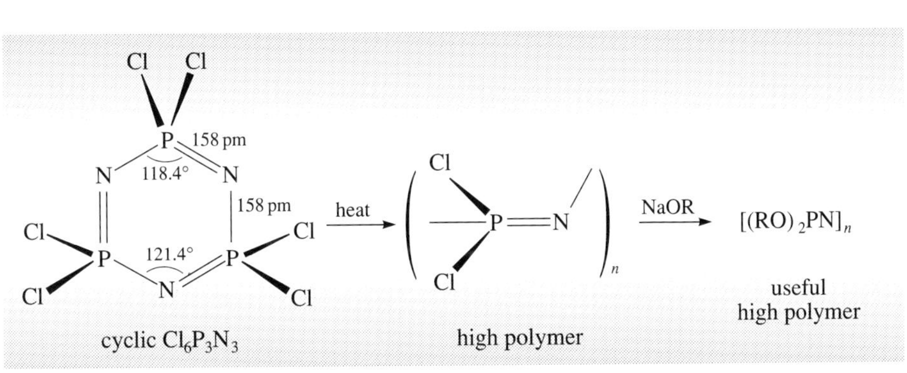

Figure 64 Structure and polymerization of the trimeric phosphazene, *cyclo*-$(Cl_2PN)_3$.

$Cl_6P_3N_3$ is made by heating together ammonium chloride, NH_4Cl, and phosphorus pentachloride, PCl_5. It can be polymerized by heating when it is very pure. It forms a high molecular mass rubbery solid, $(Cl_2PN)_n$, which is rather sensitive to water and air — not very useful! However, the P—Cl groups can be replaced by others which are more stable. Thus, if the —Cl groups are replaced by —OR groups, where —OR is a mixture of $-OCH_2CF_3$ and $-OCH_2(CF_2)_nCF_2H$, the resulting polymers, $[(RO)_2PN]_n$, have outstanding resistance to organic solvents and retain elasticity at low temperatures; they are used in the manufacture of special gaskets and hoses.

Unlike the Si—O backbone of silicones, the P—N backbone of the phosphazenes is biodegradeable: there is active research on using polyphosphazenes substituted by biologically active groups to provide slow release of drugs in the body.

5.5.8 SUMMARY OF SECTION 5.5

1 There are several allotropes of phosphorus. White phosphorus has the molecular structure, P_4, whereas the red, violet and black forms contain polymeric structures.

2 Phosphorus is able to expand the octet in its bonding and, like other third-row elements, can increase its coordination number to six.

3 The phosphorus trihalides are rather volatile compounds which contain pyramidal molecules. All the phosphorus pentahalides can be formed; the fluoride has a molecular structure, whereas the chloride and bromide both have ionic structures in the solid, $[PCl_4]^+[PCl_6]^-$ and $[PBr_4]^+Br^-$ respectively.

4 Phosphine, PH_3, is the compound analogous to ammonia, but is very reactive, inflaming spontaneously in air.

5 Reaction of phosphorus with a limited supply of sulphur yields P_4S_3, an industrially important compound used in matchheads.

6 Reaction of phosphorus with a limited supply of oxygen yields P_4O_6, with its characteristic cage structure. Further oxidation adds terminal oxygen atoms to each phosphorus atom to give P_4O_{10}, a useful drying and dehydrating agent.

7 The wet acid process produces the majority of the phosphoric acid for commercial use, by the reaction of sulphuric acid on mineral phosphates.

8 It has proved possible to make compounds containing the P=P bond by stabilizing them with a bulky group. This prevents the close approach of molecules, which is necessary for polymerization to occur.

9 Polyphosphazenes, $[(RO)_2PN]_n$ (analogous to the silicones), can be made. They have special physical properties which are just beginning to find use commercially.

5.6 OXOACIDS

There are several oxoacids of phosphorus, some of which we have already mentioned. The three most important are phosphoric acid, H_3PO_4 (also known as orthophosphoric acid), phosphorous acid, H_3PO_3, and hypophosphorous acid, H_3PO_2. Other names for these acids are listed in Table 14.

■ What is the oxidation number of the phosphorus in each of these acids?

□ The oxidation numbers are +5, +3, and +1, respectively.

In Equations 90 and 91 we saw a method of making the first two of these acids, namely by hydrolysis of the appropriate oxide.

The salts of phosphoric acid, H_3PO_4, are the common phosphates, including all phosphate minerals. The mineral apatite, calcium phosphate, $Ca_3(PO_4)_2$, and its hydrated form, hydroxyapatite, $Ca_5(PO_4)_3OH$, are the principal minerals in bones and teeth.

Some acids are **polybasic**, which means that they ionize stepwise*. For example, the most important acid of phosphorus, phosphoric acid, H_3PO_4, is a tribasic acid; that is, it undergoes three successive ionizations:

$$H_3PO_4(aq) = H^+(aq) + H_2PO_4^-(aq) \quad (97)$$

$$H_2PO_4^-(aq) = H^+(aq) + HPO_4^{2-}(aq) \quad (98)$$

$$HPO_4^{2-}(aq) = H^+(aq) + PO_4^{3-}(aq) \quad (99)$$

Because phosphoric acid is tribasic, it is possible to form three series of salts with a metal such as sodium: the dihydrogen phosphate, the hydrogen phosphate and the normal phosphate. The normal salt formed by NH_4^+, the ammonium ion, is ammonium phosphate, $(NH_4)_3PO_4$, which is important as a fertilizer.

■ What would you expect to be the salts of phosphorous acid?

□ Phosphorous acid, H_3PO_3, is dibasic and so there are two series of salts, known as phosphites. The sodium salts are NaH_2PO_3 and Na_2HPO_3.

Phosphorous acid and the phosphites are useful reducing agents. The structure of phosphorous acid is shown in Figure 65.

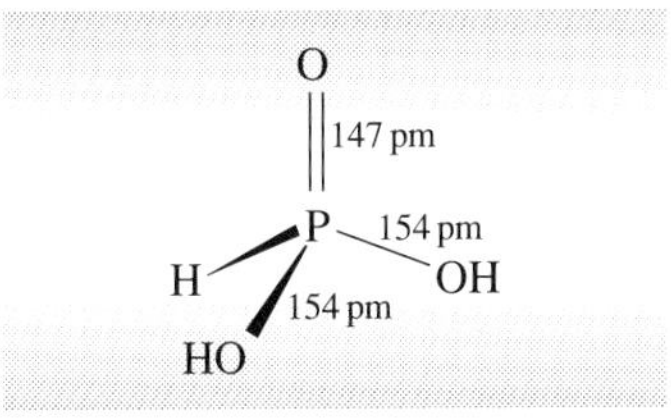

Figure 65 The structure of phosphorous acid, H_3PO_3, showing the interatomic distances found by X-ray diffraction studies.

A large variety of phosphorus acids is derived from 'polyacids', which contain two or more acidic phosphorus centres (these are considered in detail in Section 5.6.5).

The common phosphorus acids are summarized in Table 14. Both their common names and the new systematic nomenclature are given.

Table 14 Common phosphorus oxoacids and their anions

Formula	Traditional name	Traditional anion name	Hydrogen nomenclature	Structure
H_3PO_4	phosphoric acid *or* orthophosphoric acid	phosphate *or* orthophosphate	trihydrogen tetraoxophosphate (3–)	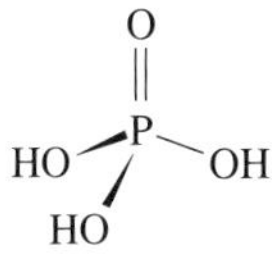
H_3PO_3	phosphorous acid *or* phosphonic acid	phosphite *or* phosphonate	dihydrogen hydridotrioxo-phosphate (2–)	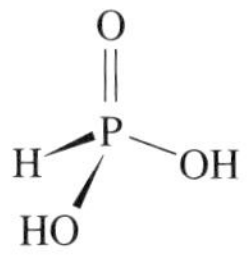
H_3PO_2	hypophosphorous acid *or* phosphinic acid	hypophosphite *or* phosphinate	hydrogen dihydrido-dioxophosphate (1–)	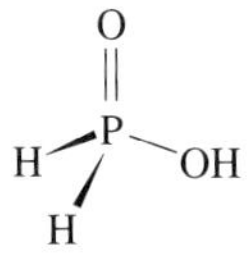
$H_4P_2O_7$	diphosphoric acid	diphosphate	tetrahydrogen μ-oxo-hexaoxodiphosphate	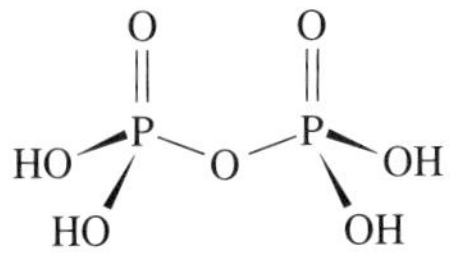
$(HPO_3)_n$ (in the limit where $n = \infty$)	metaphosphoric acid	metaphosphate	poly[hydrogen trioxophosphate (1–)]	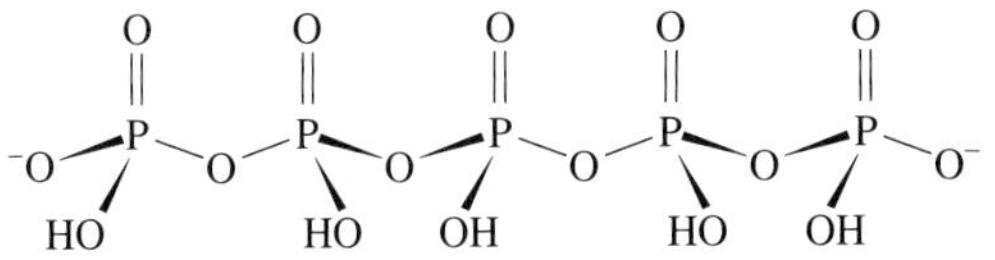

* Compare with, for example, hydrochloric acid, HCl, which is a *monobasic acid*.

Oxoacids by definition contain a covalent AO—H bond, which can dissociate to give a proton and an oxoanion:

$$AO{-}H = AO^- + H^+ \qquad (100)$$

There may in addition be one or more terminal oxygen atoms, so the **general formula of oxoacids** is $A(O)_t(OH)_n$, where t can equal 0. On this formulation, sulphuric acid, H_2SO_4, is written as $S(O)_2(OH)_2$ (where t = 2, n = 2), and boric acid as $B(OH)_3$ (where t = 0, n = 3).

■ What are the values of t and n for phosphoric acid?

□ From Figure 62 (p.72) we see that the formula can be rewritten as $P(O)(OH)_3$, giving t = 1 and n = 3.

These covalent hydroxo-compounds have available a wide range of structural possibilities, which is the reason for the existence of a relatively large number of oxoacids. The variables are as follows:

1 There may be several —OH groups in the acid, each one of which can dissociate to form a proton and an oxoanion. Consider again, for example, the three successive ionizations of phosphoric acid, H_3PO_4 :

$$\text{stage 1:} \quad H_3PO_4(aq) = H^+(aq) + H_2PO_4^-(aq) \qquad (97)$$

$$\text{stage 2:} \quad H_2PO_4^-(aq) = H^+(aq) + HPO_4^{2-}(aq) \qquad (98)$$

$$\text{stage 3:} \quad HPO_4^{2-}(aq) = H^+(aq) + PO_4^{3-}(aq) \qquad (99)$$

Each of the species on the left-hand side of the three equations is a different oxoacid, and each of the equilibria has a different dissociation constant. The equilibrium constant for dissociation of the first proton (stage 1), or 'first dissociation constant', is given the symbol K_1 (7.5×10^{-3} mol l^{-1} for this dissociation), the second dissociation constant is K_2 (6.2×10^{-8} mol l^{-1} here), and the third dissociation constant is given the symbol K_3 (1.0×10^{-12} mol l^{-1} here).

2 Oxoacids may condense* to form 'dimers, 'trimers and 'polymers'. Thus, phosphoric acid, H_3PO_4, can self-condense to produce diphosphoric acid and triphosphoric acid:

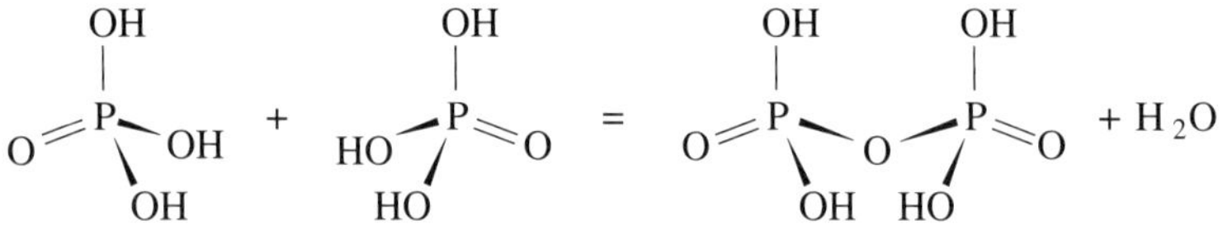

(101)

diphosphoric acid

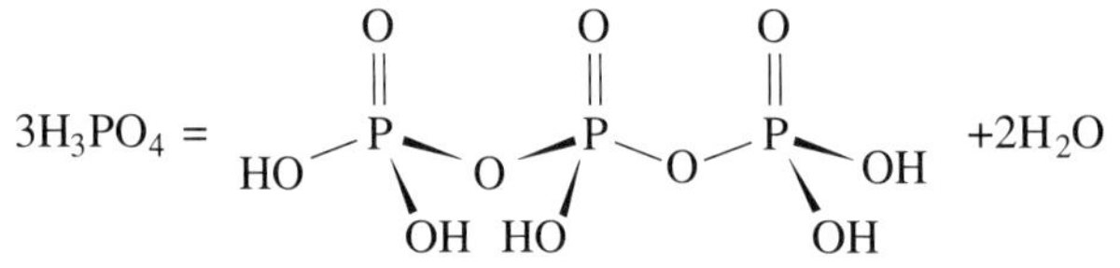

(102)

triphosphoric acid

It can also form higher polymers described by the general term 'metaphosphoric acid'.

3 The central element, A, may exist in more than one oxidation number. As we saw in Block 5, oxygen stabilizes high oxidation numbers. In fact, the range of oxidation numbers found in compounds with oxygen is wider than that with any other element except fluorine. Phosphorus, for example, forms oxoacids in oxidation numbers +5, +3 and +1. Chlorine forms the oxoacids $HClO_4$, $HClO_3$, $HClO_2$ and HClO.

* The term 'condensation' refers to reaction of two or more molecules to form a larger molecule with the elimination of a small molecule such as water; for example

$$R{-}OH + HOR' \longrightarrow R{-}OR' + H_2O$$

It is common to describe condensed oxoacids loosely as 'dimers' and 'trimers', although, strictly speaking, they are not polymers because their formulae differ from those of the monomer by one or more H_2O units.

■ What is the oxidation number of chlorine in each of these acids?

□ The halogen has, respectively, oxidation numbers of +7, +5, +3 and +1.

4 A final complication, arising with sulphur, is that this Group VI element can take the place of oxygen in an oxoacid, so that there is the possibility of one or more S—S bond, in what is then called a thio-acid. We shall meet examples of this later in the Block when we look at the chemistry of sulphur in detail.

5.6.1 OXOACID FORMULAE

The interpretation and prediction of oxoacid formulae often presents difficulty to the inexperienced chemist. This is partly because of structural complexity, but it is not helped by the irrational tradition of writing oxoacid formulae with hydrogen first, which conceals the fact that hydrogen is bonded to oxygen in many cases. (The formulae of organic oxoacids are written more sensibly, for example, CH_3COOH for ethanoic acid.) $Si(OH)_4$ is a much more helpful description of orthosilicic acid (the meaning of the prefix 'ortho' is described in Section 5.6.2) than the more commonly used H_4SiO_4, and, as we have seen, $(OH)_3PO$ is a more meaningful, but rarely used, formulation of phosphoric acid than H_3PO_4 (see Table 14). However, there seems to be no general move to rationalize the formulation of oxoacids, and their nomenclature has been to a large extent exempted from the systematization applied to other molecules. Although we shall explain the new nomenclature below, you will rarely find it used for the common inorganic acids. It is further complicated by the fact that the inorganic acids and their organic derivatives also have different common names; for instance, *phosphorous* acid in inorganic chemistry becomes *phosphonic* acid for its organic derivatives! We have tried to help you through this potential minefield by underlining prefixes and suffixes in the next Section, and always using the name in most common use.

The conventions applied at the present time to the naming of oxoacids are outlined in the next Section.

5.6.2 NOMENCLATURE OF OXOACIDS

Historically, where the central element forms oxoacids in two oxidation numbers, the higher state is indicated by the suffix -ic (as in phosphoric and sulphuric acids), and the lower state by the suffix -ous (as in phosphorous and the hypothetical sulphurous acid.)

If *more* than two oxidation numbers are involved, the prefixes per- and hypo- are used as well: per- denotes the highest oxidation number, and hypo- the lowest oxidation number. Thus, the oxoacids of chlorine are described as follows:

formula	$HClO_4$	$HClO_3$	$HClO_2$	$HClO$
oxidation number	+7	+5	+3	+1
name	perchloric	chloric	chlorous	hypochlorous

Oxoanions derived from -ic acids are given the ending -ate; examples are provided by the sulphate ion, SO_4^{2-}, phosphate ion, PO_4^{3-}, chlorate ion, ClO_3^-, perchlorate ion, ClO_4^-. Oxoanions derived from -ous acids are given the ending -ite; examples are the sulphite ion, SO_3^{2-}, nitrite ion, NO_2^-, hypochlorite ion, OCl^-, phosphite ion, PO_3^{3-}.

Condensed forms of oxoacids are also distinguished by means of prefixes. Thus, the prefix ortho- refers to the 'monomeric', or most highly hydroxylated, form, and the prefix meta- refers to the 'polymeric', or least highly hydroxylated, form.

The prefixes di- and tri- in this context are self explanatory, and refer to 'dimers' or 'trimers', as on p. 76. The oxoanions are analogously labelled. Thus, we have orthophosphates, PO_4^{3-}, diphosphates, $P_2O_7^{4-}$, and triphosphates, $P_3O_{16}^{5-}$. (Condensed anions may also be labelled with the prefix cyclo- or catena- to distinguish cyclic from linear forms, respectively.)

Table 14 also contains a column with the systematic name given to the acids under the IUPAC system. Because these names have not gained wide acceptance yet, we shall only make you aware of them. The names are derived as though the acids were salts. The name is in two parts, with first the acidic hydrogens treated separately from the rest of the compound, which then is named as a coordination complex. These names are potentially very useful because they immediately tell you how many ionizable protons there are in the acid.

5.6.3 PREDICTION OF FORMULAE

For an element in its highest oxidation number (and sometimes others), it is possible to predict the formula of its orthoacid from coordination number considerations. The condensed oxoacid formulae are then easily derived by subtracting the appropriate number of water molecules.

Third- and fourth-row elements prefer four-coordination in their oxoanions. For example, using the oxidation number approach, we can see that when phosphorus(V) coordinates four O^{2-} ions to form a complex ion of stoichiometry PO_4, the resultant charge on this oxoanion is (+5 − 8) = −3. Thus, the formula of the corresponding neutral oxoacid is H_3PO_4.

■ Predict the formula of the oxoacid fomed by sulphur(VI).

□ Sulphur(VI) coordinates four O^{2-} ions to form a complex ion of stoichiometry SO_4, with resultant charge (+6 − 8) = −2, so the oxoacid is H_2SO_4.

The chlorine(VII) oxoanion, with stoichiometry ClO_4, has charge (+7 − 8) = −1, so only one proton is required for formation of the neutral oxoacid $HClO_4$.

We should note here that, in oxoanions of the second Period, such as carbonate, CO_3^{2-}, three coordination is the norm because it allows the formation of a planar assembly with extensive π-bonding.

The above arguments often fail to predict the formulae of oxoacids of elements in lower oxidation numbers, because the preferred coordination number is either not achieved, or is achieved only by formation of a direct link between the central atom and hydrogen, as you will see in the next Section.

SAQ 24 Predict the formulae of the oxoacids of nitrogen(V) and boron(III).

5.6.4 STRENGTHS OF OXOACIDS

Strengths of oxoacids can be affected by many variables. For example, you may wonder how acidity depends on oxidation number, or how it varies across a row of the Periodic Table. There is also the question of the relationship between the first, second, etc., dissociation constants of polybasic acids.

Fortunately, two simple generalizations developed by Linus Pauling allow prediction of the acidities of oxoacids with a fair degree of accuracy.

Pauling's first rule relates to successive dissociation constants of a polybasic acid. From simple electrostatic considerations, it is easy to predict that successive ionization steps will take place less readily (for example, you might argue that it is easier to lose a proton from a neutral species like H_3PO_4 than from one, like $H_2PO_4^-$, which already carries a negative charge). The value of the rule is that it allows us to *quantify* the relationships between successive dissociation constants.

Pauling's first rule states that successive dissociation constants K_1, K_2, K_3,..., are in the ratio $1 : 10^{-5} : 10^{-10}$..., etc.

To generalize,

$$\frac{K_{n-1}}{K_n} \sim 10^5$$

This rule holds well for the common oxoacids, as Table 15 shows.

Table 15 Successive dissociation constants of H_2SO_4 and H_3PO_4

Acid	K_1/mol l^{-1}	K_2/mol l^{-1}	K_3/mol l^{-1}	K_1/K_2	K_2/K_3
H_2SO_4	$\sim 10^3$	1.2×10^{-2}		8×10^4	
H_3PO_4	7.5×10^{-3}	6.2×10^{-8}	1.0×10^{-12}	1.2×10^5	6.2×10^4

Pauling's second rule relates acid strength to the number of non-hydrogenated oxygen atoms in the molecule. The more of these there are, the stronger is the acid. The first acid dissociation constant, K_1, is predicted to be of the order of 10^{5t-8} mol l^{-1}, where t is the number of terminal (non-hydrogenated) oxygens in the acid (Block 5, Section 9.1.2).

If t in the general formula $AO_t(OH)_n$ is zero (that is, the acid is a *hydroxo* acid), the acid is very weak, with $K_1 \sim 10^{-8}$ mol l^{-1}.

If $t = 1$, the acid is moderately weak $K_1 \sim 10^{-3}$ mol l^{-1}.

If $t = 2$ or higher, the acid is strong ($K_1 \sim 10^2$ mol l^{-1} or above).

To a first approximation, the strength is independent of n, the number of —OH groups.

Pauling's second rule can be rationalized in terms of the electron-withdrawing effect of the terminal oxygen atoms. The highly electronegative oxygen atoms withdraw electron density from the —OH group, thus rendering its hydrogen more positive, and so more easily ionizable. We would expect that the more terminal oxygen atoms there are, the easier it will be to ionize the —OH hydrogen and thus the stronger will be the acid.

This second rule holds fairly well for all oxoacids, although, bearing in mind the wide variation in the nature, properties and environment of the central atom, it is not surprising that the range of K_1 values for various values of t is rather wide (Table 16).

SAQ 25 Cite two acids with $t = 0$, two with $t = 1$, two with $t = 2$ and one with $t = 3$.

SAQ 26 Predict the structures and order of acid strength for the chlorine oxoacids: $HClO$, $HClO_2$, $HClO_3$ and $HClO_4$.

SAQ 27 Use Pauling's second rule to predict how the acidity of the highest oxoacid of an element varies across the fourth Period, from Group IV to Group VII. (You should deduce the formulae of the acids by use of the oxidation number approach, as outlined in Section 5.6.3.)

SAQ 28 Arrange the following oxoacids in order of acid strength using Pauling's second rule *alone:*

H_5IO_6, $HReO_4$, H_3AsO_3, H_2CrO_4

SAQ 29 The acid of stoichiometry H_3PO_3 has a first dissociation constant of 1.6×10^{-2} mol l^{-1}. Can this be explained on the basis of Pauling's second rule? What does the K_1 value suggest about the structure?

SAQ 30 The acid of formula H_3PO_2 has a K_1 value of 1×10^{-2} mol l^{-1}. Suggest a structure for this acid.

Table 16 First dissociation constants of various oxoacids

$t = 0$		$t = 1$		$t = 2$ or more	
Formula	$K_1/\text{mol l}^{-1}$	Formula	$K_1/\text{mol l}^{-1}$	Formula	$K_1/\text{mol l}^{-1}$
HOCl	2.9×10^{-8}	$HClO_2$	1.1×10^{-2}	$HClO_3$	large
HOBr	2.1×10^{-9}	H_3PO_4	0.75×10^{-2}	H_2SO_4	$\sim 10^3$
H_4SiO_4	1.0×10^{-10}	HNO_2	0.45×10^{-2}	$HClO_4$	large
H_3BO_3	5.5×10^{-10}				

5.6.5 CONDENSATION OF ORTHOACIDS

The tendency for an orthoacid to polymerize by **condensation** is most marked in the less acidic (more highly hydroxylated) acids. There are many stable condensed forms of silicic and boric acid, whereas the condensed forms of phosphoric acid are unstable towards hydrolysis.

■ Why do Group VII oxoacids not usually yield condensed oxoacids?

□ Because they usually contain only one —OH group, so the product of condensation is an oxide. (One exception to this is periodic acid, H_5IO_6, which has five —OH groups and does form condensed oxoacids.)

Thus, dehydration of perchloric acid, $HClO_4$, gives dichlorine heptoxide, Cl_2O_7:

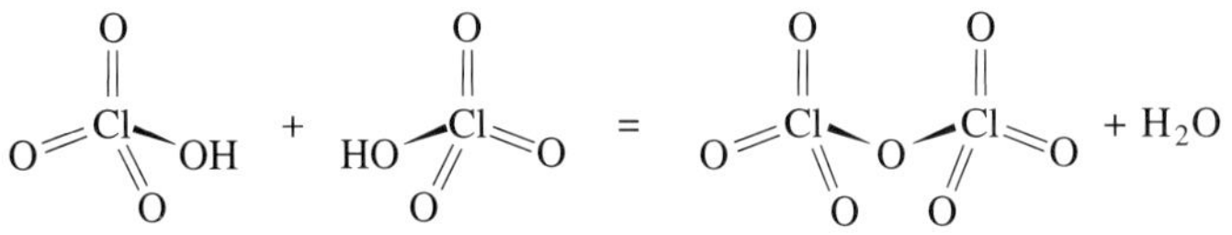

(103)

Condensation is most marked in structures where the charge on the uncondensed anion is high, because it is able to reduce the charge density on the anion. The dimerization of SiO_4^{4-} can be represented as follows:

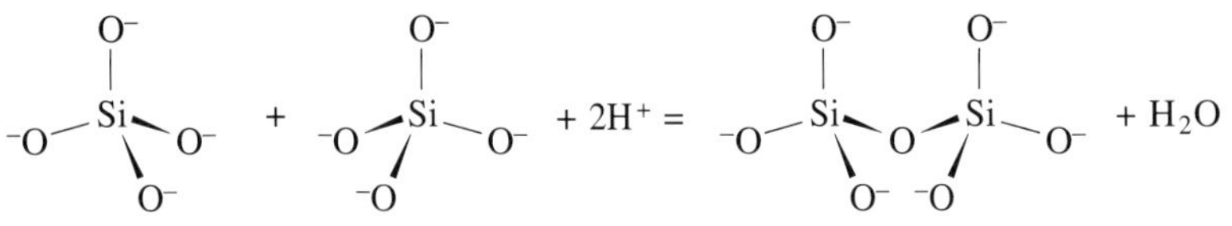

(104)

In the monomer there are four negative charges for four oxygen atoms; in the dimer there are six negative charges and seven oxygen atoms; in the trimer (Structure **29**) there are eight negative charges and ten oxygen atoms.

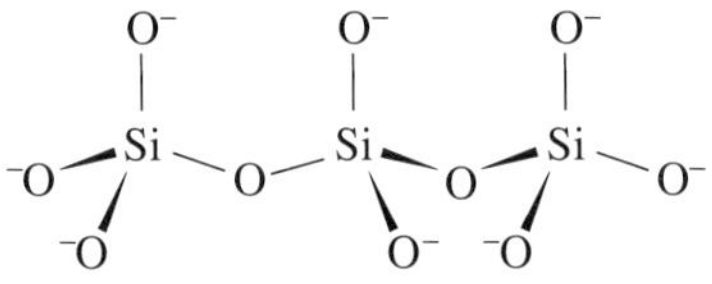

29

In the limiting case of the infinite polymer, there are two negative charges to every three oxygen atoms. This is the structural unit found in the pyroxene group of minerals. As we saw in Section 4.3.2, shared SiO_4 tetrahedra can be assembled into rings, chains, double chains, sheets and three-dimensional networks to give the amazing variety of structures found in the crystalline silicate minerals.

Phosphate tetrahedra, PO_4^{3-}, can link up via oxygen-sharing to give polyphosphates in the form of chains and rings. The phosphates, unlike the silicates, contain a central atom with valency five. Thus, the maximum number of oxygen atoms that each tetrahedron can share is *three*, since there must always be one vertex of the tetrahedron which is a terminal oxygen, bound as P=O.

■ We have already met one compound where each phosphorus forms the maximum three P—O—P linkages. Which is it?

□ Phosphoric oxide, P_4O_{10} (Figure 60b).

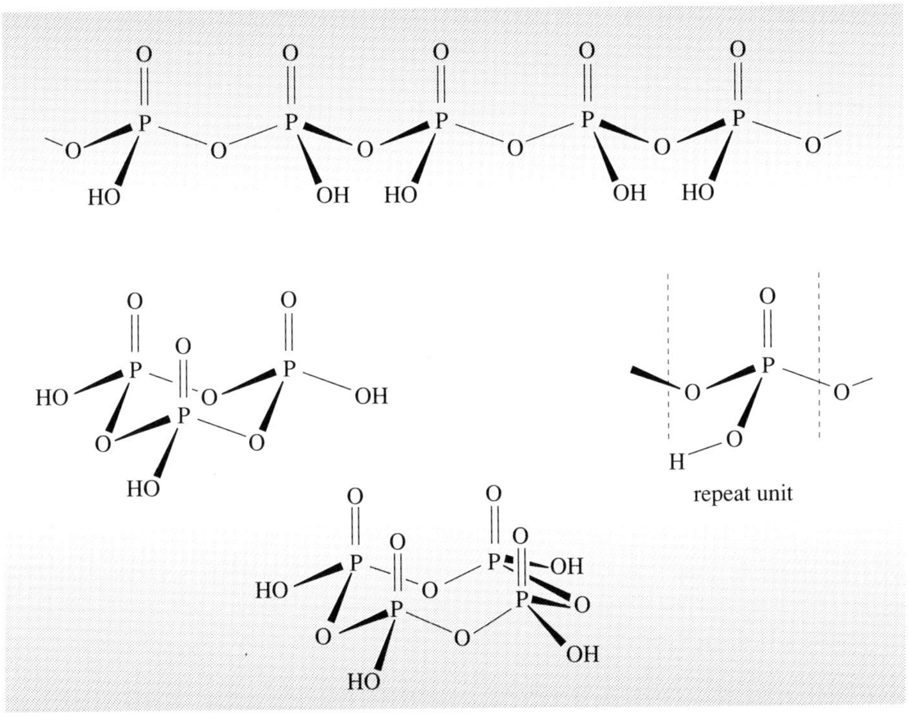

Figure 66 Metaphosphoric acid chains and rings.

Condensed phosphates can be formed if phosphoric oxide is treated with limited amounts of water; or by dehydrating phosphates by heating.

Two molecules of phosphoric acid can split off a molecule of water between them; the two tetrahedra share one oxygen, giving an acid of formula $H_4P_2O_7$, known as 'pyrophosphoric', or (more correctly) diphosphoric acid. In practice, this can be obtained by heating phosphoric acid ('pyro' comes from the Greek word for fire). Continuation of this process gives triphosphoric acid, $H_5P_3O_{10}$, and, eventually, chain polymers called metaphosphoric acids, which contain repeating $[HPO_3]$ units (Figure 66).

The ring polymers shown in Figure 66 are also called metaphosphoric acids.

The P—O—P link in polyphosphates is readily hydrolysed; in excess water, metaphosphates revert to orthophosphate. So, unlike the condensed silicates, polyphosphates are never found as minerals.

Sodium polyphosphates are the best known; the general reaction for their formation is the dehydration of sodium dihydrogen phosphate by heating. The temperature of dehydration controls the nature of the product, as indicated in Figure 67.

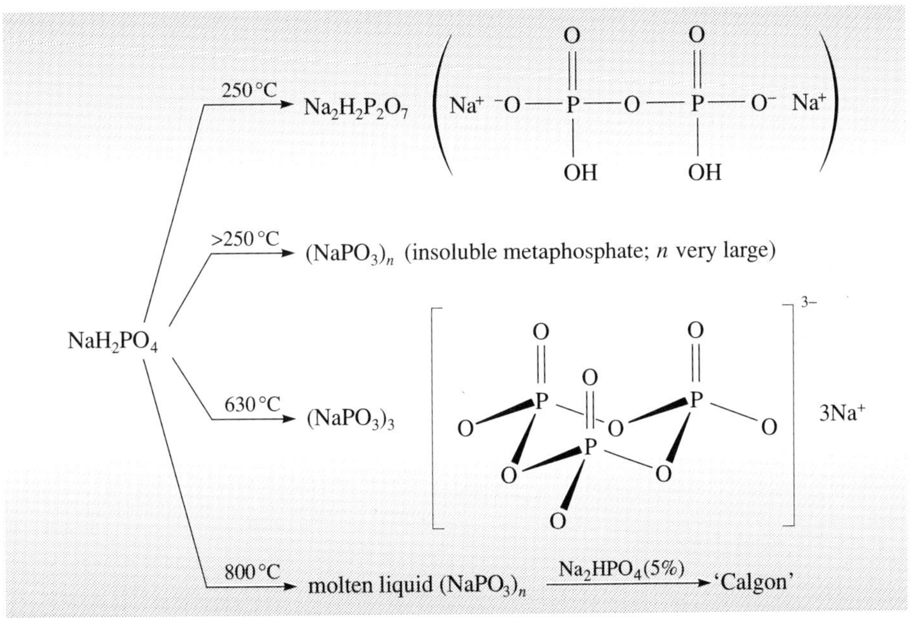

Figure 67 The dehydration of sodium dihydrogen phosphate at various temperatures.

Partial dehydration at low temperature (*c.* 250 °C) produces disodium dihydrogen diphosphate, $Na_2H_2P_2O_7$, a substance used in baking powders as a slow-acting acid for the controlled release of CO_2 gas, which produces the aerated texture of cakes, etc.

Long-chain polyphosphates are produced by heating NaH_2PO_4 above 250 °C.

The sodium salt $Na_3P_3O_9$, containing the cyclic $P_3O_9{}^{3-}$ anion, is formed when NaH_2PO_4 is heated to 600–640 °C and the melt then maintained at 500 °C.

High-temperature dehydration (800 °C) above the melting temperature of NaH_2PO_4 (628 °C) yields a liquid with very high relative molecular mass, which gives slightly soluble solids of formula $(NaPO_3)_n$:

$$nNaH_2PO_4 = (NaPO_3)_n + nH_2O \qquad (105)$$

The addition of 5% Na_2HPO_4, to provide end-of-chain units for the polymer, reduces the chain length and the product becomes soluble. This material, which is called 'Calgon' commercially, is used to *soften* water* because it can coordinate with calcium ions in solution, thereby preventing them from precipitating as insoluble $CaCO_3$ scale (Calgon = calcium gone!) The structure of Calgon is shown in Figure 68.

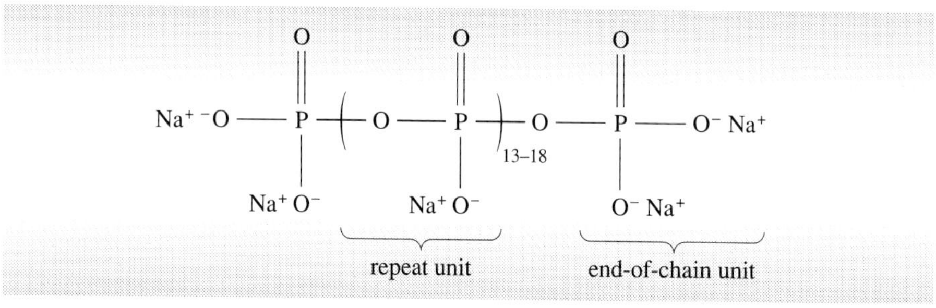

Figure 68 The structure of 'Calgon'.

Heating together a 2 : 1 mixture of Na_2HPO_4 and NaH_2PO_4 at 450 °C produces a chain of only three phosphate units:

$$2Na_2HPO_4 + NaH_2PO_4 = Na_5P_3O_{10} + 2H_2O \qquad (106)$$

The structure of the triphosphate ion, $P_3O_{10}{}^{5-}$, is shown in Figure 69.

The triphosphate ion is also very good at complexing calcium ions and holding them in solution. The coordination of a metal ion to a ligand can have a profound effect on the properties of the metal.

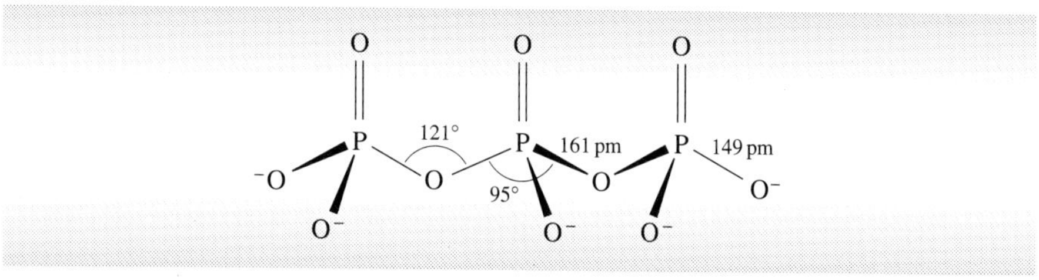

Figure 69 The structure of the triphosphate ion, $P_3O_{10}{}^{5-}$.

* Hard water contains significant concentrations of calcium and magnesium salts. The calcium and magnesium are precipitated as a scum of salts of the long-chain carboxylic acids used in soaps and detergents. In 'temporary' hard water the principal anion is HCO_3^-(aq). This form of hardness can be removed by boiling. The calcium (and magnesium) is deposited as a limestone scale, observed, for example, as the 'furring' of kettles. 'Permanent' hardness is due to high concentrations of sulphate; it is not susceptible to removal by boiling, but can be treated by Calgon.

Pentasodium triphosphate (known industrially as sodium tripolyphosphate, STPP) is added to detergents in substantial quantities. Most detergents comprise long molecules with **hydrophobic** (water-hating) and **hydrophilic** (water-loving) sections, as indicated in the example shown in Structure **30**. The basic action is for the organic hydrophobic section to bury itself in the dirt, and the hydrophilic section then allows the insoluble dirt to 'dissolve' in the water as shown in Figure 70. Sodium tripolyphosphate is able to complex with calcium and magnesium ions, Ca^{2+} and Mg^{2+}, thus softening the water and preventing these ions from reacting with the detergent and rendering it inactive. It also keeps the wash water slightly alkaline, because it is the salt of a fairly weak acid.

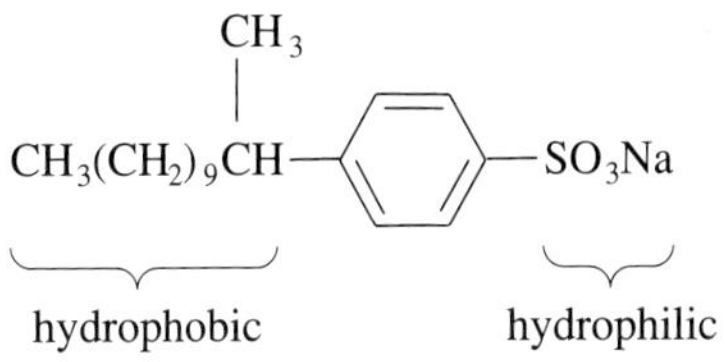

30

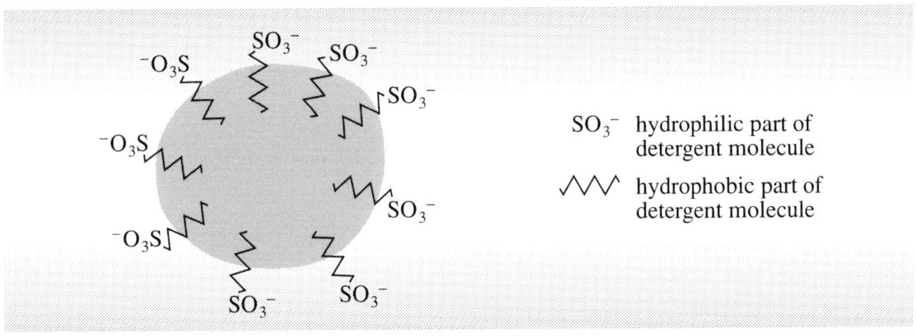

Figure 70 Grease droplet rendered 'soluble' in water by detergent.

SAQ 31 Why does calcium precipitate from hard water and how does Calgon prevent this from happening?

SAQ 32 In the Science Foundation Course Summer School, you studied the interconversion of chromate, CrO_4^{2-}, and dichromate, $Cr_2O_7^{2-}$ ions. Derive structural formulae for the parent acids of these salts. What can you say about the tendency of chromic acid to condense? Would you expect metachromates to be formed?

5.7 PHOSPHATE FERTILIZERS AND EUTROPHICATION

There is a vast number of domestic and industrial uses of phosphorus — far too many to cover in any detail here. Over a hundred million tonnes of phosphate rock are mined each year, and about 90% of this rock is used to produce impure phosphoric acid, which is mainly used to make fertilizers. The other 10% is converted into phosphorus (Figure 71), which is then used to make all the other phosphorus compounds used by the food, detergent, match and metallurgical industries which we summarized in Figure 57.

Figure 71 Industrial use of phosphorus.

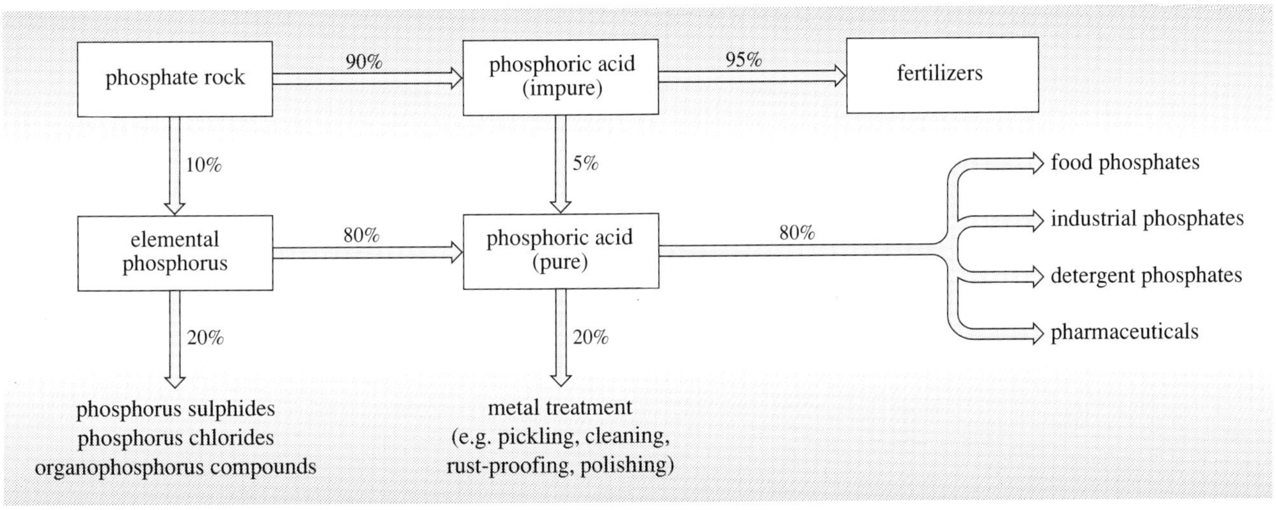

Plants require phosphorus to produce a good root system and to make seeds. Why does the phosphate rock have to be converted to phosphoric acid before it can be made into a useful fertilizer? As the phosphate mineral fluoroapatite, ($Ca_5(PO_4)_3F$), is highly insoluble, it is useless to spread it directly on the ground; plants cannot assimilate phosphorus in this form, so it needs to be converted to a more-soluble compound. However, a fertilizer must not be too soluble or it would rapidly be leached from the soil. Like fluoroapatite, calcium phosphate, $Ca_3(PO_4)_2$ is also not very soluble. However, monocalcium dihydrogen phosphate, $Ca(H_2PO_4)_2$, is moderately soluble and makes a good fertilizer. In its most useful form this is made by treating calcium phosphate with phosphoric acid:

$$Ca_3(PO_4)_2(s) + 4H_3PO_4(aq) + 3H_2O(l) = 3Ca(H_2PO_4)_2.H_2O(s) \quad (107)$$

when the product is known as triple superphosphate. $Ca(H_2PO_4)_2$ can also be made by treating fluoroapatite with sulphuric acid:

$$2Ca_5(PO_4)_3F(s) + 7H_2SO_4(aq) + 3H_2O(l) = 7CaSO_4(s) + 3Ca(H_2PO_4)_2.H_2O(s) + 2HF(g) \quad (108)$$

The mixed product of monocalcium dihydrogen phosphate and calcium sulphate is marketed as superphosphate. In this form small amounts of fluoride remain, which may be unacceptable where soils already have a high fluoride content. Moreover, the calcium sulphate does not have any nutritive value for the soil; it acts as an inert diluent to the fertilizer.

The phosphoric acid for the triple superphosphate process is made using the wet acid process, where the phosphate rock (formulated as calcium phosphate in Equation 93) is reacted with sulphuric acid. Phosphates play an essential role in modern society, particularly in the production of fertilizers and detergents. It has been found in recent years, however, that the large amount of phosphate used in developed countries is not without its problems. Inevitably not all the phosphate fertilizer is taken up by plants; some is leached out of the soil and finds its way into the rivers and lakes. Added to this are the phosphates finding their way into the sewerage system from detergents and industrial processes. Because phosphates are nutrients, their presence in river and lake waters have an effect on the growth of plants and algae.

Plants and algae need many nutrients as well as heat and light in order to grow. A simple picture is that a particular plant species will grow as long as all the necessary nutrients are available, but that when any one of the nutrients needed is used up, growth stops. The nutrient that is in short supply is said to be **limiting**. It is not uncommon for phosphorus to be the limiting nutrient.

Imagine an ecologically balanced lake in which algal growth is limited by the supply of phosphates. What happens if this supply is increased by, say, the expansion of a lakeside town or a change in farming methods? The first effect would be an increase in algal growth. This, however, would not necessarily continue in line with the increased phosphate supply. It may be that a different nutrient then becomes limiting, and further increases in phosphate would have no effect on growth. Again, growth of algae could continue until a balance was achieved by the grazing of fish. It is important to realize that an increase in the supply of a specific nutrient can only bring about an increased growth provided that the nutrient is limiting.

At certain times of the year, large blooms of algae may appear on the surface of polluted inland waters: superficially, they can seem to be merely an inconvenience, making swimming or sailing less pleasant, but it is more serious than this. As the weather gets colder, the growth slows down and finally stops. The algae will then sink and decay. The absorbed nutrients in the algae return to the water, but dissolved oxygen is consumed in the decay process. This can result not only in an unpleasantly smelling lake (largely due to the production of hydrogen sulphide), but in an extreme case the fish in the lake can die, due to the lack of oxygen in the water; the lake is then said to be **eutrophic**. Plate 9 shows algal growth in eutrophic water. It does not always happen that large algal growth causes a lake to die, but smaller growths can cause changes in fish population. Such results have a great effect on the biological food chain and can greatly change the ecology of a lake.

Problems of this type have not been too serious in the UK, the main places affected being Lough Neagh in Northern Ireland and the Norfolk Broads, but in the USA there have been acute problems with some of the Great Lakes.

5.8 ARSENIC, ANTIMONY AND BISMUTH

The halides of arsenic, antimony and bismuth illustrate the following trends down the Group; (a) increasing metallic character of the elements; (b) the inert pair effect — that is, the tendency towards an increased stability of the 'Group number minus two' oxidation number (+3); (c) tendency to higher coordination numbers.

All three pentafluorides, AsF_5, SbF_5 and BiF_5, are known. The only other pentahalide stable at room temperature is $SbCl_5$; $AsCl_5$ is only stable at low temperatures. A demonstration of the tendency to higher coordination numbers is given by the pentafluorides. Arsenic pentafluoride is a monomeric gas. At normal temperatures, antimony pentafluoride is a viscous liquid containing six-coordinate antimony in *cis* fluorine-bridged polymers (Figure 72a). It crystallizes to give a structure that contains *cis*-bridged tetramers (Figure 72b). Bismuth pentafluoride is a white crystalline solid, in which the bismuth coordination number is six, but in this case the structure has *trans* fluorine bridges, giving infinite chains (Figure 73). All these fluorides are highly reactive, hydrolysing rapidly in air to give hydrofluoric acid. They have to be handled with great care!

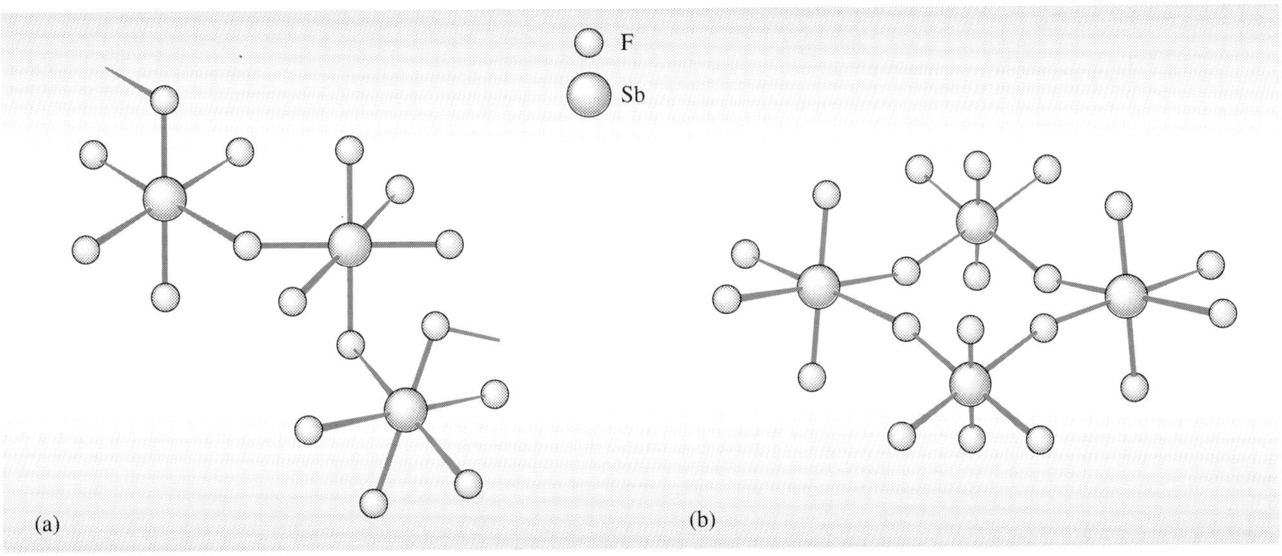

Figure 72 (a) The structure of liquid SbF_5; (b) the structure of crystalline SbF_5.

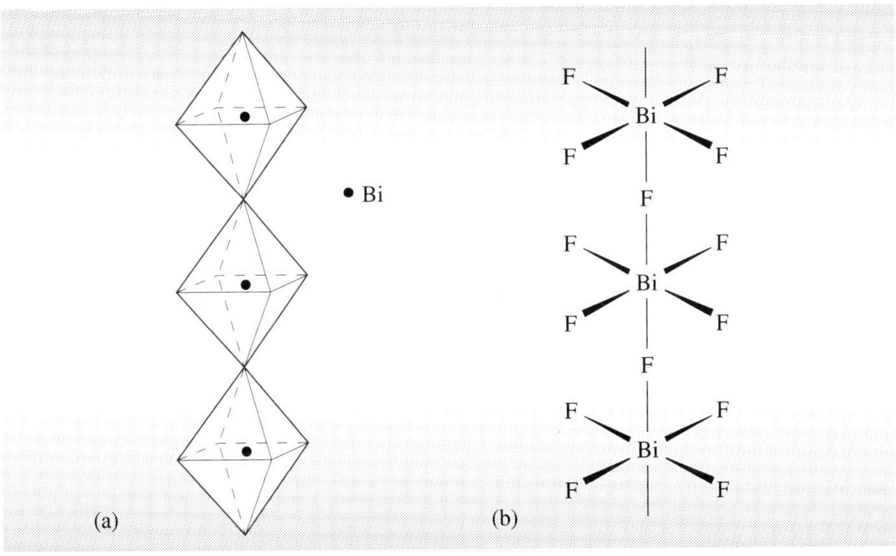

Figure 73 The structure of crystalline BiF_5: (a) corner-linked BiF_6 octahedra; (b) structural formula.

All the trihalides, MX_3 (M = As, Sb, Bi and X = F, Cl, Br, I) are known. They are mostly volatile and are rapidly hydrolysed by water.

When bismuth is dissolved in molten $BiCl_3$, a black solid of formula $Bi_{24}Cl_{28}$ is obtained. This contains some very interesting ionic species: $[BiCl_5]^{2-}$, $[Bi_2Cl_8]^{2-}$ and $[Bi_9]^{5+}$, in the ratio 4 : 1 : 2. The ion $[Bi_9]^{5+}$ consists of the metal atom cluster shown in Figure 74.

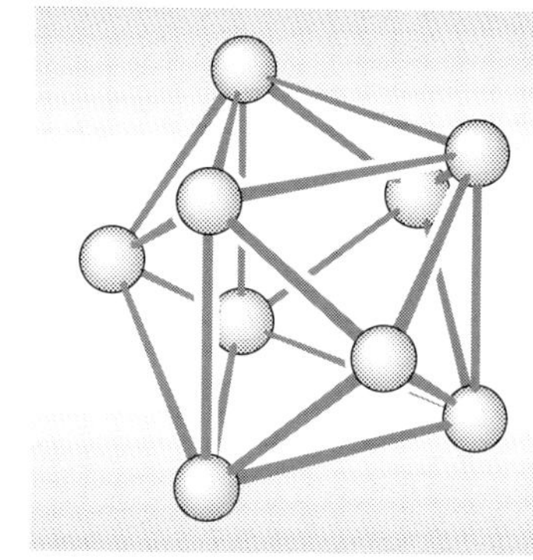

Figure 74 The structure of the $[Bi_9]^{5+}$ ion, in which a trigonal prism of bismuth atoms has three others outside (capping) the rectangular faces.

Arsenic(III) and antimony(III) both form oxides of formula X_4O_6. The molecules have the same structure as P_4O_6 in the gas phase. Like P_4O_6, arsenic(III) oxide is acidic; antimony(III) oxide is amphoteric. Bismuth forms a basic oxide, Bi_2O_3, which is a yellow powder at room temperature. The pentoxides of these elements are less well characterized, Bi_2O_5 being extremely unstable to loss of oxygen. Compounds containing E=E (E = As, Sb and Bi) bonds have been synthesised. As with the corresponding phosphorus compounds, stabilization of the double bond is achieved by substitution with bulky groups such as tertiary butyl.

SAQ 33 How many different types of fluorine atoms are there in liquid SbF_5?

5.9 SUMMARY OF SECTIONS 5.6–5.8

1 Oxoacids of elements in Groups III–VII may exist for several oxidation numbers of the central element, and as monomers (ortho oxoacids) or condensation polymers (meta oxoacids).

2 Ortho oxoacids have a general formula $AO_t(OH)_n$, which may be predicted from coordination number and oxidation number considerations.

3 Acid strength is higher for dissociation of the first proton in a polybasic acid than for the second (and so on).

4 Acid strength increases as the value of t increases.

5 The relative strengths of oxoacids can be predicted from Pauling's rules:

$$\frac{K_{n-1}}{K_n} \sim 10^5 \text{ and } K_1 \sim 10^{5t-8}$$

If the number of non-hydrogenated oxygen atoms, $t = 0$, the acid is weak, $K_1 \sim 10^{-8}\,\text{mol l}^{-1}$; if $t = 1$, the acid is moderately strong, $K_1 \sim 10^{-3}\,\text{mol l}^{-1}$; if $t = 2$, the acid is strong, $K_1 \sim 10^{2}\,\text{mol l}^{-1}$ or more.

6 The tendency to condensation is highest with weak acids. Thus, orthosilicic acid readily converts to a meta form, which does not happen for sulphuric or perchloric acids.

7 Monocalcium dihydrogen phosphate (triple superphosphate), $Ca(H_2PO_4)_2$, is made as a fertilizer from calcium phosphate and phosphoric acid, or by treating the mineral fluoroapatite with sulphuric acid, when it is known as 'superphosphate'. Excessive run-off of phosphates into lakes can cause algal growth and eutrophication.

8 The chemistry of arsenic, antimony and bismuth illustrate trends down a Group such as increasing metallic character, the inert pair effect and the tendency to higher coordination numbers.

6 THE GROUP VI ELEMENTS

6.1 STRUCTURES AND PROPERTIES OF THE ELEMENTS

Group VI (Figure 75) is the last of the Groups we shall be studying in this Block. Like Groups III, IV and V, it includes non-metals (oxygen, sulphur and selenium), a semi-metal (tellurium), and a metal (polonium). Once again you should notice an increase in metallic properties in descending a Group in the Periodic Table.

Electronic configurations of the Group VI elements are shown in Table 17.

Oxygen is the most abundant element on our planet. It forms 46% by mass of the Earth's crust, much of it occurring in silicates; it comprises 23% of the atmosphere, where it occurs as the gaseous element, O_2; and it forms about 85% of the hydrosphere, where it is combined with hydrogen as water, H_2O.

Oxygen is generated naturally in photosynthesis, whereby plants synthesise carbohydrates, etc., from water and carbon dioxide in the atmosphere, using sunlight as the source of energy. The overall equation for the reaction is:

$$6CO_2(g) + 6H_2O(l) \xrightarrow[\text{chlorophyll}]{h\nu} 6O_2(g) + \underset{\text{carbohydrates, etc.}}{C_6H_{12}O_6} \qquad (109)$$

Oxygen is necessary for both plant and animal life. It is taken in during the respiration process and carbon dioxide is breathed out, so completing the cycle.

Oxygen is a colourless, odourless gas. It has three stable isotopes, ^{16}O, ^{17}O, and ^{18}O, of which ^{16}O is by far the most abundant (99.76%, 0.04% and 0.2%, respectively).

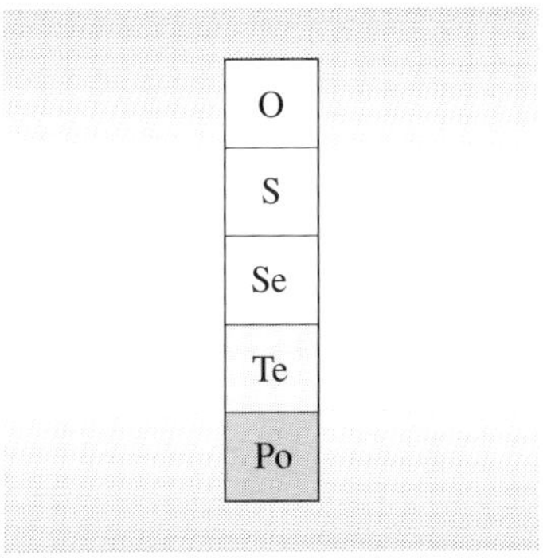

Figure 75 The Group VI elements: the metal (polonium) is shaded dark grey; the semi-metal (tellurium) is shaded light grey; non-metals are unshaded.

Table 17 Electronic configurations of Group VI atoms

Atom	Electronic configuration
O	$[He]2s^22p^4$
S	$[Ne]3s^23p^4$
Se	$[Ar]3d^{10}4s^24p^4$
Te	$[Kr]4d^{10}5s^25p^4$
Po	$[Xe]4f^{14}5d^{10}6s^26p^4$

■ What is the magnetism of the oxygen molecule?

□ O_2 is paramagnetic; it has two unpaired electrons in the $2p\pi_g$ level.

(If you cannot remember the molecular orbital diagram, it is a good idea to look back now to Block 4, Figure 57.) This property was first noticed by Michael Faraday (Figure 76) in 1848. Oxygen gas was first liquefied in 1877; it condenses to a pale blue paramagnetic liquid at −180.0 °C and atmospheric pressure.

Figure 76 Michael Faraday, born 1791 in Newington, Surrey; died 1867. Faraday was the son of a blacksmith. He left school at 13 and was apprenticed to a bookseller and bookbinder. He began his scientific experiments after reading the entry on electricity in a volume of *Encyclopaedia Britannica* which he was binding. After attending lectures by Humphry Davy at the Royal Institution, he eventually became his assistant. He stayed at the RI for the rest of his working life, becoming Professor in 1833. He had many brilliant achievements, including the liquefaction of chlorine and the discovery of benzene; he discovered electromagnetic induction and devised the first electric motor; he showed that a magnet moving in and out of a coil of wire induced a current, and formulated the laws of electrolysis. He was a very private man and could be a non-conformer: he refused a knighthood and the Presidency of the Royal Society saying 'I have always felt that there is something degrading in offering rewards for intellectual exertion, and that societies or academies or even kings and emperors should mingle in the matter does not remove the degradation'. In 1826 he started the famous Christmas lectures for children at the RI, which are still popular today and are broadcast on BBC television.

Ozone, O_3, is a naturally occurring allotrope of oxygen. It has a characteristic strong smell, which the human nose can detect in concentrations as low as 0.01 p.p.m.! Indeed it was originally named after the Greek, οζειν, meaning smell.

■ What do you predict to be the shape of O_3?

□ The Lewis structure for O_3 has the two resonance forms represented by Structures **31** and **32**, and Structures **33** and **34**. The lone pair on the central oxygen leads to a bent molecule.

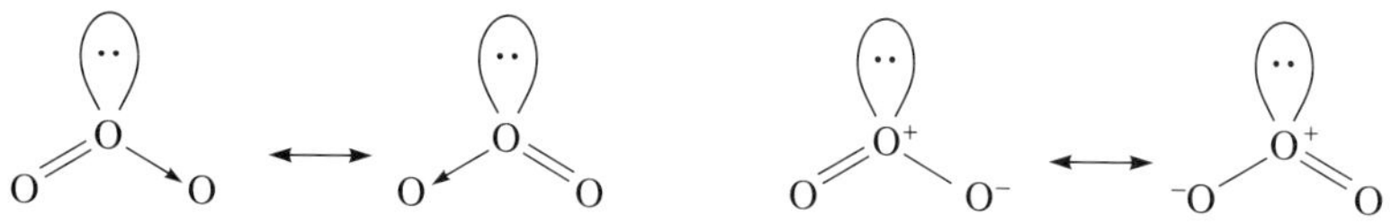

31 **32** **33** **34**

Remember that we cannot describe O_3 as in Structure **35**, because this would involve expanding the octet of a second-row element. Microwave spectroscopy of O_3 indicates a bond angle of 116.8° and an O—O bond distance of 127.8 pm: this compares with the O—O distance of 120.7 pm in doubly bonded O_2 and 149 pm in the singly bonded peroxide ion, O_2^{2-}.

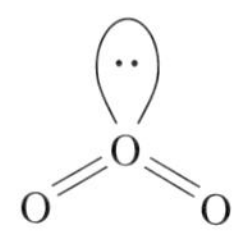

35

Because of its dramatic associations with volcanic eruptions, sulphur has been recognized since antiquity; the first reference to the element as 'brimstone' was in the Book of Genesis. As early as the third century AD, Chinese alchemists made use of sulphur mixed with saltpetre (KNO_3) as a primitive gunpowder. They also valued it as one of the constituents of the 'pill of immortality', which was said to have been responsible for the untimely deaths of no less than seven Chinese emperors. In fact, the blame for this early pharmaceutical failure rests not with sulphur, but with another favoured constituent, which was mercury!

To alchemists of all traditions, sulphur was held to be one of the basic ingredients of matter; it was the principle of fire, and believed for a long time to be a constituent of anything combustible. However, in the eighteenth century, sulphur was one of the first substances to be firmly established as an element in the modern sense, by the work of the French chemist Antoine Lavoisier. It is interesting to note that, long before these theoretical questions were settled, sulphur chemistry was already being exploited by early technologists. The commercial preparation of sulphuric acid (a process of vital importance to contemporary chemical industry) goes back to the fourteenth century.

Sulphur is not very abundant in the Solar System, but tends to concentrate (in association with iron) in the cores of the inner (iron-rich) planets. The Earth's core is thought to contain as much as 15% by mass of sulphur, but this source is not, of course, accessible to us. However, the 0.05% in the Earth's crust seems to be ample for projected current needs. Sulphur is found as sulphide ores combined with metals, such as galena, PbS, iron pyrites, FeS_2 (known as 'fool's gold'; see Plate 10), cinnabar, HgS, and zinc blende, ZnS; removal of sulphur from these ores (whether the sulphur is recovered or emitted into the atmosphere) is a necessary accompaniment to the metal extraction process. There are widespread deposits of elemental sulphur located in areas of high volcanic activity such as Japan and the Andes, or found in association with petroleum. The latter 'sedimentary' sulphur deposits result from the reducing action of bacteria on sulphate in sediments during the process of rock formation. (Sulphate deposits form a large proportion of world reserves of sulphur, but, as we are not yet able to convert sulphate efficiently to sulphur in a low-cost process (unlike bacteria!), these deposits are not, at present, commercially useful.)

Petroleum-linked sulphur deposits often occur at some depth below the surface, and it was not until the beginning of the twentieth century, when the elegant Frasch recovery method (devised by Hermann Frasch) was developed, that they became commercially important.

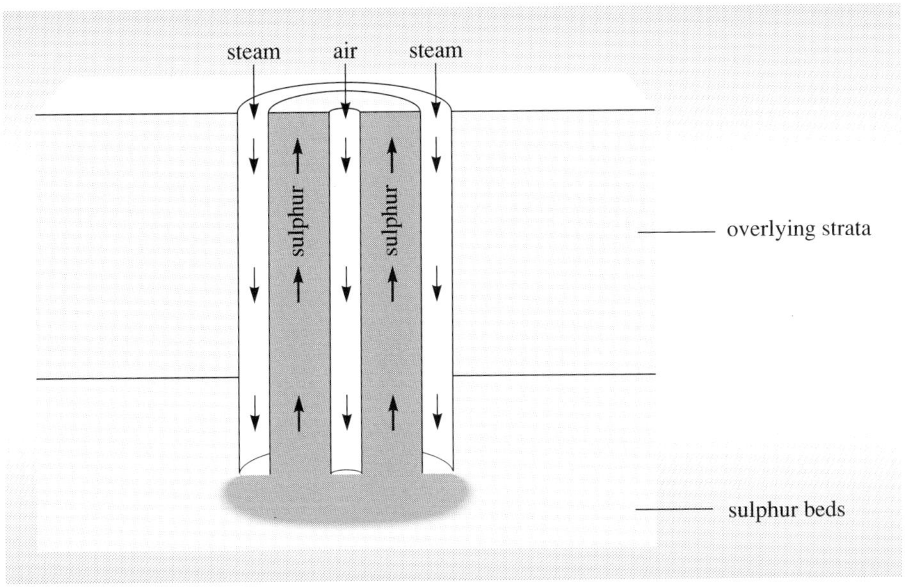

Figure 77 The Frasch sulphur extraction process.

In the **Frasch process** (Figure 77), three coaxial tubes are inserted in a shaft of about 30 cm diameter. High-pressure steam at 165 °C is forced down the outer pipe to melt the elemental sulphur, which is then blown up through the third pipe with the assistance of compressed air passed down the centre pipe. The beauty of this process is that it encompasses in a single stage both mining and purification. The high-quality product (99% pure sulphur) is thus obtained relatively cheaply.

All fossil fuels — gas, coal and oil — contain sulphur, as did the plants from which they originated. The combustion of these fuels, if untreated, is therefore certain to produce sulphur dioxide, SO_2, as a by-product. Even in low concentration, sulphur dioxide is harmful to vegetation, and the adverse environmental effects of SO_2 (or rather, of the products of its oxidation in polluted urban air) are discussed in Case Study 4 and Videocassette 2, Sequence 14. The law requires SO_2 emission to be minimized, so oil and gas are routinely purified to remove sulphur before combustion. Much of the sulphur used commercially today derives from this source.

It is a much more difficult matter to remove sulphur from solids such as coal prior to combustion, although dense metal sulphides like iron pyrites, FeS_2, can be partly removed by slurrying techniques. Removal of sulphur dioxide after combustion is difficult to achieve at economical costs. There is often about 1% of SO_2 in the combustion gases from burning coal for electrical power production. This can be reacted in wet or dry processes with calcium or magnesium hydroxide or oxide, in a process known as *scrubbing*. The newest facilities remove 90–95% of the SO_2, but the cost of the electricity produced rises by several per cent. Sulphur dioxide emission from a 2 000 MW power station then falls from about 30 tonnes a day without scrubbing to about 2.5 tonnes with scrubbing. This is a very worthwhile improvement but it still constitutes a major source of atmospheric pollution.

Sulphur is a bright yellow solid, soluble in organic solvents, from which it may be recrystallized easily. The form stable at room temperature is called **orthorhombic sulphur** because of the crystalline structure adopted (Plate 11). The unit cell contains crown-shaped S_8 molecules (Figure 78), stacked in a complex array. The only difference between orthorhombic sulphur and the high-temperature (above the transition temperature of 96 °C) monoclinic modification, with which it may readily be interconverted, is merely a difference in the stacking pattern of the crowns. At room temperature the monoclinic form slowly changes into the orthorhombic form.

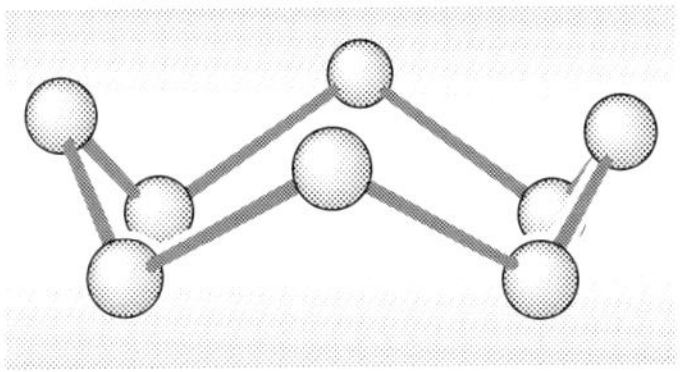

Figure 78 Octasulphur, crown form.

It is possible to make modifications that do not contain octasulphur rings — ring sizes from S_5 to S_{20} are known — but all revert in time to the more stable S_8 form. By thermal decomposition of S_2I_2 or by careful decomposition of sodium thiosulphate,

$Na_2S_2O_3$, with acid, chair-shaped S_6 rings can be prepared (Figure 79a). S_7 rings (Figure 79b) are prepared by melting sulphur at 200 °C, cooling to 159 °C and pouring into liquid nitrogen (−196 °C); the resulting solid is then extracted with a non-polar organic solvent.

We might ask, 'why do the rings revert in time to the most-stable S_8 form?' If you think of the bonding around a sulphur atom in a chain of sulphur atoms, two of the six valence electrons are used to make bonding pairs with the adjacent sulphurs, leaving two lone pairs of electrons.

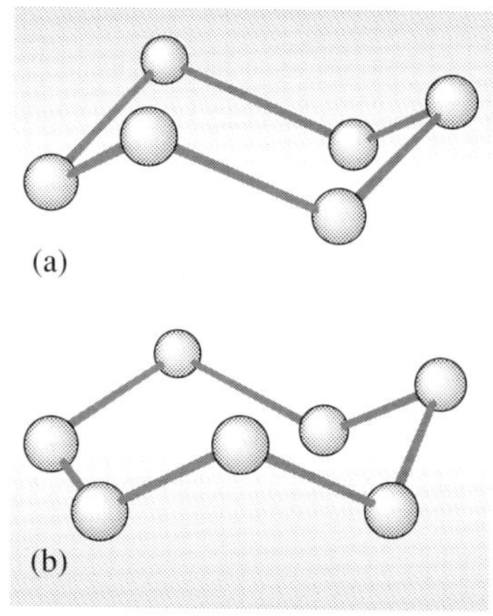

Figure 79 (a) Hexasulphur in the chair form; (b) heptasulphur.

■ How will the electron pairs arrange themselves?

□ The electron pairs will want to be as far apart as possible and to give minimum interference with electrons on the neighbouring sulphur atoms. With effectively four pairs of electrons, an approximately tetrahedral arrangement will be preferred.

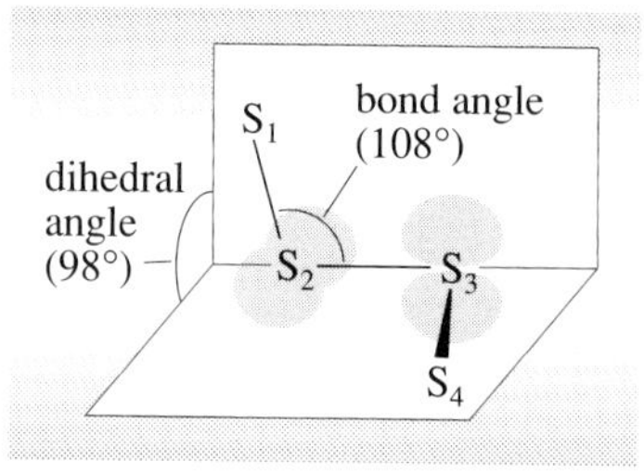

Figure 80 Angles involved in a sequence of four S atoms in the S_8 molecule.

Look at the chain of four sulphur atoms in Figure 80 taken from an S_8 ring. We find that the SSS bond angle is 108°, and the **dihedral angle** (the angle between the two planes) is 98°. In this arrangement the lone pairs are as far away from each other as possible. All other forms of sulphur have less ideal dihedral angles: in S_6 the bond angles are 102° and the dihedral angles are 74°; in the more-strained structure of S_7, the bond angles range from 101°–107°, with the dihedral angles varying between almost zero and 108°.

Orthorhombic sulphur melts to a yellow liquid at 112 °C. As the temperature of the liquid is raised above about 160 °C, its colour changes to orange, brown and almost black as the S_8 rings open to form **diradicals** (that is, radical species in which there are two atoms that have an unpaired electron), •S—(S_6)—S•. These join together to form chains of varying length. The darkening is due to the presence of the unpaired electrons on the chain ends (recall the F-centres in the alkali halides — Block 2, Section 8.1.1). The colour change is associated with a corresponding increase of viscosity. The average chain length at 170 °C is about 10^6 sulphur atoms, decreasing markedly at higher temperatures. When liquid sulphur at 160 °C is cooled rapidly by pouring into cold water, a dark rubbery polymeric solid called **plastic sulphur** results (Plates 12a and b). The solid consists of randomly orientated helices, with about three atoms to a turn. This polymer has potentially useful mechanical properties. It can be stretched to 20 times its original length, when it can form fibres with tensile strength comparable to that of Nylon. Unfortunately, the conversion to the brittle orthorhombic sulphur takes less than a month at normal temperatures.

The vapour of sulphur, at temperatures below 500 °C and at atmospheric pressure, consists mainly of S_8 rings, but, as the temperature is increased and pressure reduced, the dark violet paramagnetic substance S_2 becomes the major constituent of the vapour.

An important use of sulphur is the addition of sulphur to rubber in the **vulcanization process**. This process, which involves heating rubber with sulphur to induce cross-linkage between the long hydrocarbon chains of which rubber is composed, results in a product with superior mechanical properties.

However, the major uses of sulphur are in the heavy chemical industry. Nearly 90% of all the sulphur used (Figure 81) goes to the production of sulphuric acid, and half of this acid is used in the preparation of phosphate fertilizers. Sulphur is not incorporated in the product; sulphuric acid is chosen merely as a cheap and convenient source of protons, which are required for the process in which $Ca_3(PO_4)_2$ is converted to $Ca(H_2PO_4)_2$ (triple superphosphate). The greater solubility of $Ca(H_2PO_4)$ versus $Ca_3(PO_4)$ makes the former an effective fertilizer because the phosphorus becomes more easily assimilable by plants.

SAQ 34 How would you explain paramagnetism in a molecule of formula S_2?

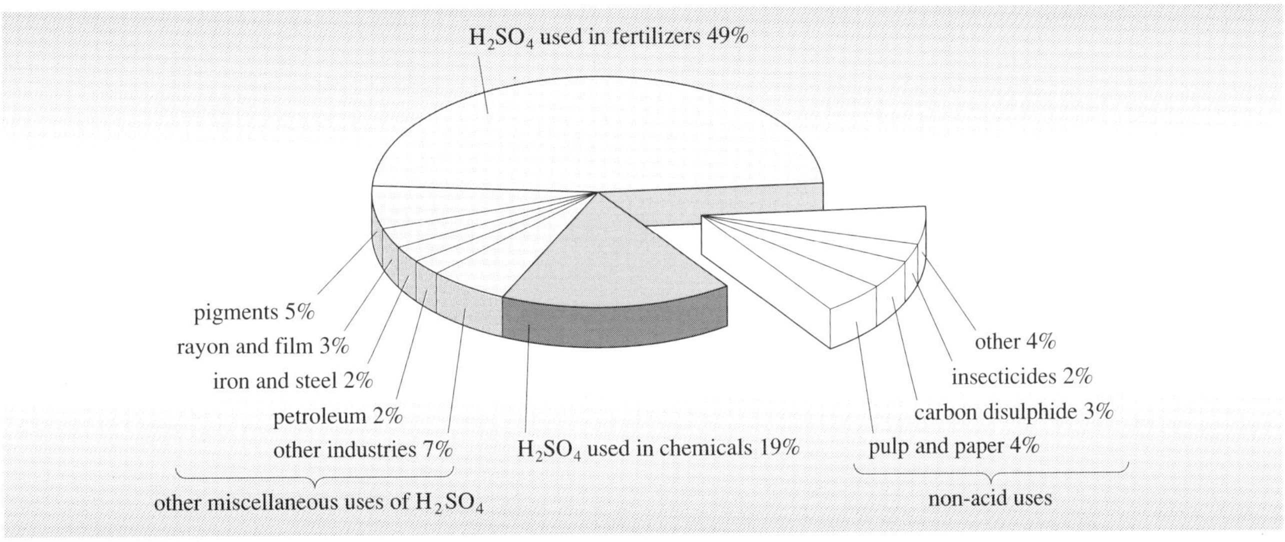

Figure 81 Uses of sulphur.

Selenium and tellurium are metallic-like solids, though selenium is classified as a non-metal and tellurium as a semi-metal (Figure 75); both elements are intrinsic semi-conductors. They both have spiral polymeric structures, although it is possible to form red allotropic forms of selenium, containing Se_8 molecules, by rapid quenching of the melt; there is no analogous molecular form of tellurium.

Polonium is a radioactive metal, the only one to have a primitive cubic structure. Polonium was first discovered by Marie Curie (Figure 82) when she managed to separate small quantities of it from tonnes of pitchblende. It was named after Poland, her home country. This work together with the parallel isolation of radium, earned her the Nobel Prize for Chemistry in 1911.

■ What is the coordination around each polonium atom in the solid?

□ In a primitive cubic structure each polonium atom is surrounded by six others at the corners of an octahedron (Block 2).

Figure 82 Maria Sklowdowska-Curie (Marie Curie; born 1867 Warsaw, Poland; died 1934) in her laboratory. Marie was a brilliant student, but her family was poor due to repression by the Russian authorities. She became a governess but dreamed of France as the land of liberty. Once in Paris she obtained a scholarship to study physics. She met and married Pierre Curie, and they set out to study the newly discovered X-rays and the radioactivity of uranium. In 1903, together with Henri Becquerel, they were awarded the Nobel Prize for Physics for this work. She went on to isolate pure radium and determine its relative atomic mass, and also discovered polonium. They refused to take out any patents on the production of radium, despite its apparent medical uses, deliberately renouncing immense financial rewards. Pierre was killed in a street accident in 1906, and when she refused a widow's pension, preferring to continue working, she was given his chair at the Sorbonne. In 1911 she was awarded a second Nobel Prize, this time for chemistry, for the discovery of radium and polonium. She had two daughters, Eve, who studied music and literature, and Irène, who followed in her parents' footsteps, eventually receiving a Nobel Prize herself in 1935. Marie Curie died in 1934 of leukaemia, a result of working with high-energy radiation all her life.

6.2 OXYGEN

Oxygen
I am the very air
you breathe
Your first
and last
breath
I welcomed you
at birth
Shall bid
farewell
at death
I am the Kiss of Life
Its ebb and flow
With your last gasp
You will call my name:
'o o o o o o o o o o o o'

Roger McGough

The bond dissociation energy of the oxygen molecule is quite high (498 kJ mol^{-1}, cf. N_2, 945 kJ mol^{-1}), but oxygen is nevertheless a very reactive gas, which oxidizes many of the elements directly (for example, hydrogen is oxidized to water).

■ Why is the bond dissociation energy high for the oxygen molecule?

□ O_2 has a double bond due to both σ and π overlap of the 2p orbitals.

Certain reactions of molecular oxygen are ubiquitous and important — respiration and combustion, for example. When an oxidation reaction gives out a lot of heat and proceeds spontaneously, we call this **combustion**. Oxygen willingly supports combustion partly because the molecule has two unpaired electrons. The oxygen molecule is a diradical, and is therefore very reactive. In the combustion process a flame or spark initiates the reaction by breaking some bonds to form some free atoms or radicals, and these start chain reactions in which radicals (or atoms) regenerate radicals (or atoms). The reaction gives out heat and light, and can be explosive because, as well as being exothermic, it goes very fast. This provides a mechanism for the strong O=O bond to react, whereby radicals interact with the unpaired electrons to open up the double bond, which then allows the weak O—O single bond to react.

6.2.1 PEROXIDES

Hydrogen peroxide, H_2O_2, is the other hydride of oxygen, apart from water. Its structure was introduced in Block 5, Section 7.2, and is shown in Figure 83.

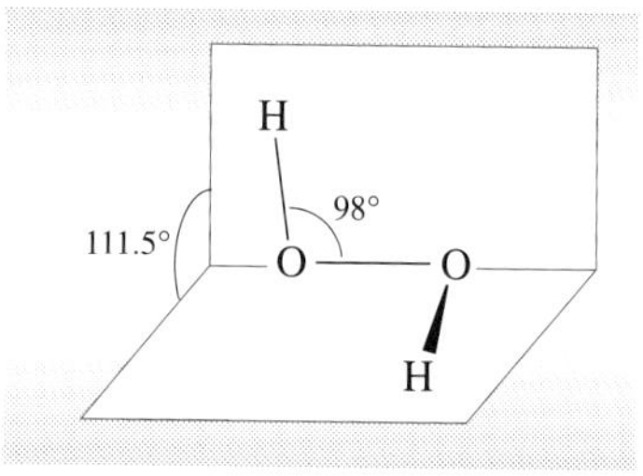

Figure 83 The most-stable conformation of the hydrogen peroxide molecule, H_2O_2, in the gas phase. The dihedral angle is 111.5°.

Peroxide salts such as Na_2O_2 and BaO_2 are formed when alkali metals and alkaline earth metals are heated in air (Block 3, Section 9.2). The action of dilute acids on these salts produces hydrogen peroxide; for example

$$BaO_2(s) + 2H^+(aq) + SO_4^{2-}(aq) = BaSO_4(s) + H_2O_2(aq) \quad (110)$$

The reaction in Equation 110 is a particularly useful method of preparation of H_2O_2, because, unusually for the salt of a Group I or II metal, barium sulphate is extremely insoluble, which shifts the equilibrium to the right.

Pure H_2O_2 and concentrated solutions in water decompose as follows:

$$H_2O_2(l) = H_2O(l) + \tfrac{1}{2}O_2(g); \qquad \Delta G_m^\ominus = -116.7\,\text{kJ mol}^{-1} \quad (111)$$

As you saw in Section 5.3.2, hydrogen peroxide solution is a strong oxidizing agent, and a weak reducing agent. Most of the hydrogen peroxide produced is used for bleaching and water purification.

6.2.2 OXIDES AND THE PERIODIC TABLE

Except for the lighter noble gases, all the elements form oxides (Table 18). Typical metals form ionic oxides, which are usually basic. Those of the alkaline earth metals (MO) have the sodium chloride structure. Non-metals form covalent oxides, which give acids with water and/or salts with alkali.

Table 18 Oxides of the typical elements

	H					
Li	Be	B	C	N	O	F
Na	Mg	Al	Si	P	S	Cl
K	Ca	Ga	Ge	As	Se	Br
Rb	Sr	In	Sn	Sb	Te	I
Cs	Ba	Tl	Pb	Bi	Po	At
	basic		amphoteric		acidic	

Certain oxides are covalent but are neither acidic nor alkaline; examples we have seen in this Block are CO, N_2O and ozone, O_3. Such oxides are called **neutral oxides**.

Some oxides can form glasses, notably B_2O_3 and SiO_2; Al_2O_3 is refractory, with an extended structure, which is largely covalent and single bonded.

As we have seen, when an element forms two or more oxides, their acidity increases with the oxidation number of the element.

The solid lines in Table 18 enclose elements with amphoteric oxides (the higher oxide, if more than one is formed); the coloured lines enclose elements with amphoteric lower oxides.

Oxygen resembles fluorine in many of its properties, for example, in its electronegativity and its ability to stabilize high oxidation numbers. The highest oxidation numbers of the elements are found in their oxides and fluorides.

6.2.3 SUMMARY OF SECTION 6.2

1 Oxygen is a reactive, gaseous, diatomic molecule constituting approximately 23% of the air we breathe.

2 It has a strong double bond due to the σ and π overlap of the 2p orbitals.

3 It is paramagnetic due to two unpaired electrons in the $2p\pi_g$ antibonding orbital.

4 Ozone, O_3, an allotropic form of oxygen, is a bent triatomic molecule.

5 Hydrogen peroxide, H_2O_2, is a strong oxidizing agent and weak reducing agent. It forms metal salts containing the peroxide ion, O_2^{2-}.

6 Apart from the lighter noble gases, all elements form oxides. Generally speaking, basic oxides (such as Na_2O) are formed by the typical metals on the left side of the Periodic Table, and acidic oxides (such as SO_3) by the non-metals on the right. Amphoteric oxides (such as Al_2O_3) fall in the middle. A few oxides, such as CO, N_2O and O_3, do not fall into these categories, and are said to be neutral.

7 Oxygen, like fluorine, tends to stabilize higher oxidation numbers.

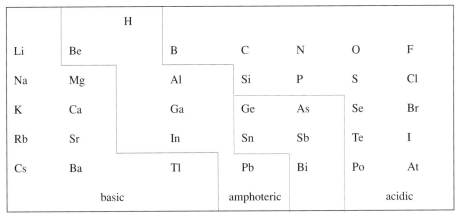

6.3 SULPHUR

Sulphur
I'm what gets witches
a bad name
funny smells
Gobbledy spells
Given to theatrics
I go in for special effects:
Brimstone and treacle
Hellfire. Eureka!
Gold! The Elixir of Life!
Immortality!
Chinese alchemists were obsessed.
Emperors were impressed
But in Beijing
I couldn't stop them
— ageing
And so they died
(But not in vain)
For a potion more mundane
was chanced upon.
The Chinese called it:
'Fire Drug'
Mobsters
got where they got with it ...
Children
play a lot with it...
Cities
glow white hot with it...
Guy Fawkes
hatched a plot with it...
Gunpowder.

Roger McGough

Sulphur, in lower oxidation numbers, resembles oxygen in its chemistry but differs from oxygen in that higher oxidation numbers such as +4 and +6 are also available. A brief examination of the compounds of sulphur with hydrogen, halogens and oxygen should serve to illustrate the similarities and differences.

6.3.1 SULPHUR HYDRIDES (SULPHANES)

Higher oxidation numbers are stabilized, in general, by oxygen and fluorine; however, compounds of hydrogen with sulphur are limited to those in which sulphur is divalent.

The most important hydride, H_2S, an important constituent of many types of natural gas, is smelly and is as poisonous as hydrogen cyanide, causing death at 100 p.p.m. The fact that H_2S is a gas (b.t. −60 °C), whereas its oxygen analogue, H_2O, is a liquid at room temperature, is a consequence of the stronger hydrogen-bonding between molecules in H_2O. Like water, however, the molecule is V-shaped. The strength of the S—H bond is lower than that of the O—H bond, as a comparison of bond enthalpy terms shows: B(O—H), 464 kJ mol^{-1}; B(S—H), 364 kJ mol^{-1}.

The bond angle in H_2S (Structure **36**) is found to be very close to 90°, which can be taken to indicate that sulphur makes use of almost pure p orbitals in its bonds to hydrogen. H_2S is a weak acid (first acid dissociation constant 10^{-7} mol l^{-1}), which is formed when acid is added to its metal salts. The gas used to be generated in the laboratory in a Kipp's apparatus, in which dilute hydrochloric acid was dropped on to iron(II) sulphide:

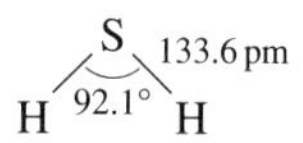

36

$$FeS(s) + 2H^+(aq) = Fe^{2+}(aq) + H_2S(g) \qquad (112)$$

If acid is added to polysulphides (see Section 6.3.2), a mixture of acids containing S—S bonds results: H_2S_2, H_2S_3, H_2S_4, etc. As these are analogues of carbon and silicon hydrides, they are called **sulphanes**. Sulphur has the highest tendency after carbon to form bonds to itself, that is, to catenate. Remember that in Block 5 (Section 7.2) we calculated the possibility of oxygen forming a catenated molecule O_8, analogous to S_8. We saw that the reason this does not happen is because the O—O single bond is so weak and B(O=O) > 2B(O—O). For sulphur the reverse is true: now the π–π overlap is poorer and the S—S single bond is stronger; B(S=S) << 2B(S—S), and so sulphur prefers catenation.

SAQ 35 Use the bond enthalpy approach to predict an approximate standard enthalpy change for Reaction 113:

$$2H_2S(g) + S_8(s) = 2H_2S_5(l) \qquad (113)$$

(H_2S_5 has the structure H—S—S—S—S—S—H.) Comment on the entropy change for this reaction. What conclusions can you draw about the stability of H_2S_5 on the basis of your calculated value of $\Delta H_m^\ominus$ and the sign of the entropy change?

SAQ 35 shows that reactions analogous to Reaction 113 are unlikely to be exothermic. Because S_8 exists as a catenated molecule, there is no change in the number of H—S and S—S bonds during the reactions involved in their preparation. Consequently, $\Delta H_m^\ominus$ will be small. However, as Reaction 113 involves a decrease in the number of moles of gas, $\Delta S_m^\ominus$ is negative, and given that $\Delta H_m^\ominus \approx 0$, the reverse reaction will be favourable, especially if T is increased. Accordingly, the higher sulphanes are unstable with respect to S_8 and H_2S, and they must be stored at low temperature to prevent the deposition of sulphur.

None of the sulphanes is highly stable with respect to the elements; even H_2S, the most stable in this respect, has a $\Delta G_m^\ominus$ of only $-33.4\,kJ\,mol^{-1}$. The stepwise decomposition of H_2S has been suggested as a route to hydrogen for use in refineries.

SAQ 36 At what temperature would a single-stage decomposition of H_2S operate? Assume the following decomposition reaction:

$$H_2S(g) = H_2(g) + \tfrac{1}{2}S_2(g) \qquad (114)$$

Use the *Data Book* for any thermodynamic data you need.

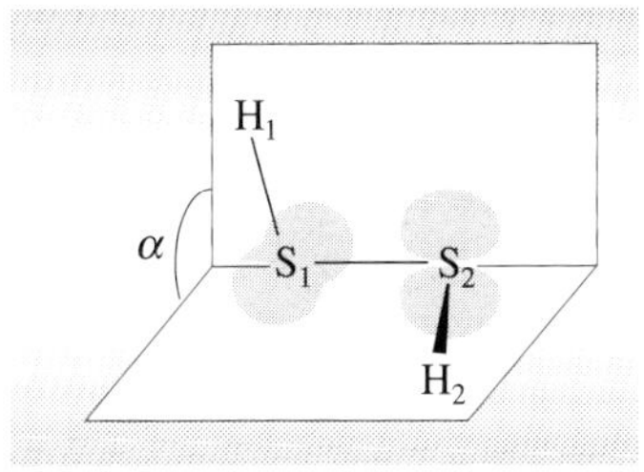

Figure 84 The dihedral angle α in H_2S_2.

The disulphane H_2S_2 (Figure 84) has a molecular structure similar to hydrogen peroxide, demonstrating the steric influence of the sulphur lone pairs. The dihedral angle in this case is the angle between the $H_1S_1S_2$ and $H_2S_2S_1$ planes.

6.3.2 SULPHIDES

Most metals react directly with sulphur to form sulphides, which in the main are very insoluble in water. The sulphides of metals from Groups I and II are ionic, and soluble in water. However, like oxides, they are salts of a weak acid, and are therefore extensively hydrolysed in solution:

$$S^{2-}(aq) + H_2O(l) = OH^-(aq) + HS^-(aq) \qquad (115)$$

Solutions of Groups I and II metal sulphides can dissolve sulphur to form anions containing sulphur chains of type S_n^{2-} (n up to 6), which are known as **polysulphides**, for example calcium pentasulphide, CaS_5. The disulphide ion, $^-S{-}S^-$, is found naturally in the metal ore, iron pyrites, FeS_2 (Plate 10).

As discussed in Block 3, SAQ 9, sodium–sulphur cells are presently being developed for use as storage batteries suitable for electrically driven vehicles. Successful prototypes have been built, but at the time of writing (1994) they are not commercially viable.

Transition-metal sulphides are less ionic in nature; some have useful semimetallic properties, such as photoconductivity, and are used, for example, in photocells. Cadmium sulphide has been used in solar cells; thin-film CdS cells were chosen in preference to silicon for solar panels on unmanned space flights in the late 1970s. Zinc and cadmium sulphides are commonly used as detectors in spectrophotometers, particularly in the infrared.

The mineral lapis lazuli gains its intense blue colour from the presence of the trisulphide ion, S_3^{2-}. These stones have been used both for decorative purposes and as pigments in paint and ceramics since ancient times.

6.3.3 SULPHUR HALIDES (HALOSULPHANES)

Sulphur uses a wide range of oxidation numbers in its compounds with halogens, particularly with fluorine. The sulphur halides (also known as *halosulphanes*) that have been characterized are listed in Table 19. Investigation of the structure of these halides presents an opportunity to practise the skills you acquired in Blocks 4 and 6.

Table 19 Sulphur halides

Oxidation number	*	+1	+2	+4	+5	+6
		S_2F_2	$[SF_2]^{\dagger}$	SF_4	S_2F_{10}	SF_6
	S_nCl_2	S_2Cl_2	SCl_2	SCl_4		
	S_nBr_2	S_2Br_2				
		S_2I_2				

* Terminal sulphur atoms in the chain have oxidation number +1; others have oxidation number 0.

† SF_2 rapidly disproportionates to sulphur and SF_4. However, we include it because its structure and dipole moment are known from spectroscopic studies on the gas at low pressure.

The product of direct combination of sulphur with fluorine is the octahedral sulphur hexafluoride, SF_6 (Structure **37**).

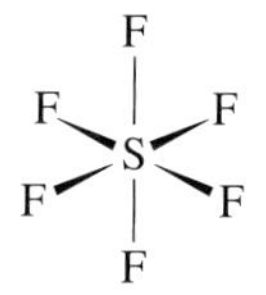

37

This capacity of fluorine to stabilize high oxidation numbers in its chemical partners is not limited to sulphur. You have met it before, for example in the interhalogen compounds (Block 5). Fluorine is the only halogen capable of oxidizing sulphur to the +6 state.

Sulphur hexafluoride is chemically unreactive. It is both the most inert sulphur compound and the most inert covalent fluoride. This is in contrast to the lower fluorides and the other sulphur halides, which are reactive and, in particular, are readily hydrolysed. The reluctance of SF_6 to take part in chemical reaction is purely kinetic; reactions such as the hydrolysis

$$SF_6(g) + 3H_2O(l) = SO_3(g) + 6HF(g); \qquad \Delta G_m^{\ominus} = -518.5\,\text{kJ mol}^{-1} \qquad (116)$$

are highly favoured thermodynamically. An obvious parallel can be found in the kinetic stability of the carbon tetrachloride/water system (SAQ 17, Section 4.3.1). The lack of a low-energy pathway for these reactions may be due to the fact that the central atom has achieved its maximum coordination number (it is said to be **coordinatively saturated**) and is protected from attack by a shell of relatively non-polarizable halogen atoms.

The lower fluorides S_2F_2 and SF_4 disproportionate readily into sulphur and SF_6. The difluoro compound, S_2F_2, demonstrates unusual structural isomerism in that it can exist in two forms: disulphur difluoride, F—S—S—F (Structure **38**) and thiothionyl fluoride, S=SF_2 (Structure **39**).

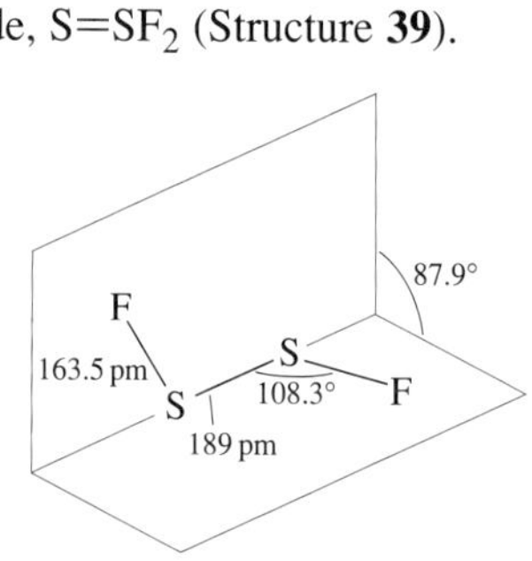

38

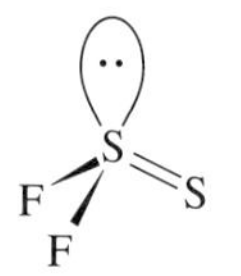

39

SF_4, with its characteristic shape (Structure **40**), is a useful selective fluorinating agent, which is capable, for example, of replacing the >C=O group in organic compounds

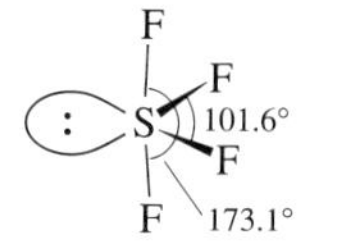

40

by $\rangle CF_2$ without affecting other functional groups. The vibrational spectroscopy of SF_4 was discussed in Block 6, Section 4.6. SF_2 is also unusual in that it exists both as the monomer and as a dimer (Structures **41** and **42**).

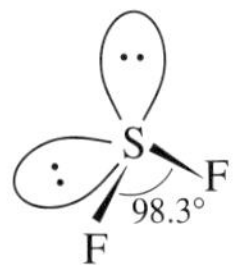

41

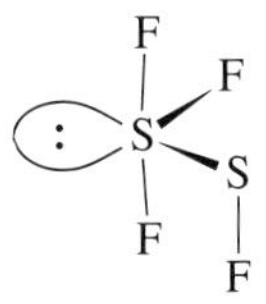

42

The normal product of direct combination of chlorine and bromine with sulphur is the so-called monohalide S_2X_2. Chlorine does not stabilize oxidation number +6 in sulphur; nor indeed is it very successful at stabilizing the +4 state (SCl_4 is stable only at low temperatures, decomposing at −31 °C into SCl_2 and chlorine). However, unlike bromine, it does stabilize the +2 state, and some dichloride, SCl_2, can be extracted from the reaction mixture if S_2Cl_2 is treated with a large excess of chlorine. Bromine forms only disulphurdibromide, S_2Br_2, and the bromosulphanes, S_nBr_2. Chloro- and bromosulphanes are most conveniently made by treating sulphanes with the monohalide; the terminal hydrogens are replaced in this reaction by —SCl or —SBr groups, increasing the sulphur chain length by 2. Chloro- and bromosulphanes with up to eight sulphur atoms in the chain have been characterized.

Figure 85 illustrates the various possible shapes for a molecule of formula S_2Cl_2, where the linkage is Cl—S—S—Cl.

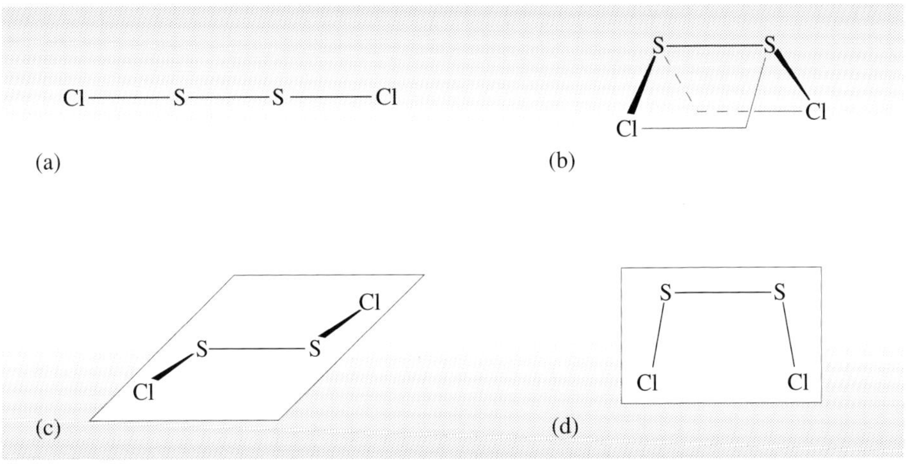

Figure 85 Structural possibilities for S_2Cl_2: (a) linear S_2Cl_2; (b) non-planar S_2Cl_2; (c) *trans*-planar S_2Cl_2; (d) *cis*-planar S_2Cl_2.

The easiest way to visualize the bent structures (b), (c) and (d) is to imagine the S—S bond to lie along the spine of a book, and the chlorine atoms to lie at opposite corners. Then, in (b) the book is considered to be partly open, in (c) to be fully flat open, and in (d) to be shut.

ACTIVITY You can try this for yourself, drawing out the structure on a sheet of stout transparent plastic, folded in two. Now try to identify the symmetry elements for each shape and then assign them to a point group. ■

The symmetry elements and point groups are identified on Figure 86.

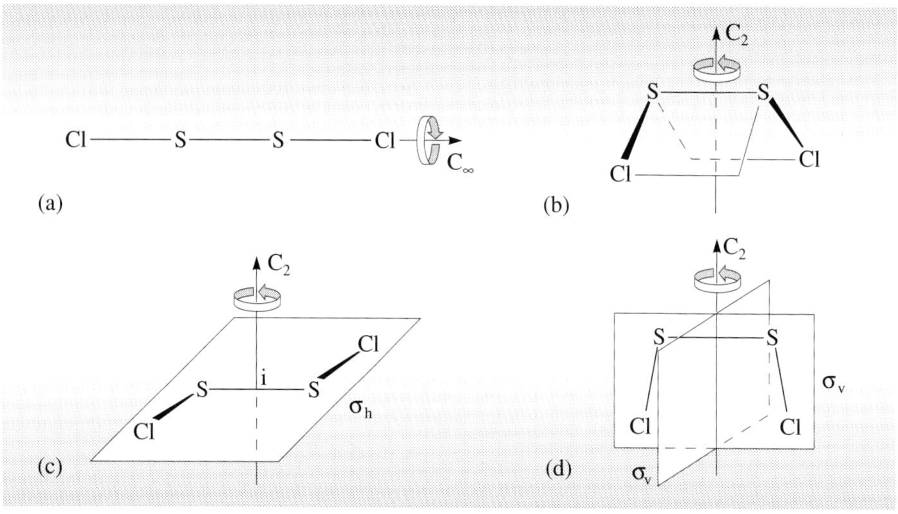

Figure 86 Symmetry elements and possible structures for S_2Cl_2: (a) linear ($\mathbf{D_{\infty h}}$) S_2Cl_2; (b) non-planar ($\mathbf{C_2}$) S_2Cl_2; (c) *trans*-planar ($\mathbf{C_{2h}}$) S_2Cl_2 ; (d) *cis*-planar ($\mathbf{C_{2v}}$) S_2Cl_2.

The linear form (a) is centrosymmetric, and thus belongs to $\mathbf{D_{\infty h}}$.

The non-planar form (b) has only one symmetry element, the twofold rotation axis, C_2. Accordingly, it belongs to the $\mathbf{C_2}$ point group.

The *cis*-planar form (d) has, in addition, two planes of symmetry, both of which contain the rotation axis. This classifies both as σ_v-type planes, so the point group of (d) is $\mathbf{C_{2v}}$.

In the *trans*-planar form (c) the molecular plane is perpendicular to the rotation axis, so that it is correctly described as a σ_h plane. Form (c) also has a centre of symmetry, i, and therefore belongs to the $\mathbf{C_{2h}}$ point group.

S_2I_2 is unstable, but can be prepared by shaking aqueous KI with S_2Cl_2 in carbon disulphide: it decomposes to S_8 and other even-numbered ring allotropes.

SAQ 37 Apply the arguments of Block 6 to answer the following questions:

(a) In which of the symmetry point groups indicated for S_2Cl_2 in Figure 86 do you expect to observe the ν(S—S) stretching mode in the infrared?

(b) How many ν(S—Cl) modes do you expect to observe in the infrared for each of the structures?

(c) Is it possible to distinguish between all four structural possibilities on the basis of the number of *stretching* modes observed in the infrared and Raman spectra?

SAQ 38 An alternative structure for S_2X_2 molecules is the $S{=}SX_2$ structure, as found in S_2F_2. On the basis of VSEPR theory, what symmetry point group would you expect this molecule to belong to? Would you expect to observe the sulphur–sulphur stretch in the infrared?

How many ν(S—X) modes do you expect to see in the infrared? How might you hope to distinguish this molecule from *cis*-planar or non-planar S_2X_2 by investigation of its stretching modes in the infrared and Raman?

6.3.4 OXIDES OF SULPHUR

The sulphur–oxygen bond is an important feature of sulphur chemistry, just as the silicon–oxygen bond is of silicon chemistry. Not only are there the simple oxides described below, and oxoanions derived from the oxoacids, but there are important oxohalides and oxonitrides which we cannot include in this brief account.

The best-known simple oxides are the dioxide, SO_2, and the trioxide, SO_3 (although highly reactive lower oxides SO and S_2O have been described). The bonding in SO_2

and SO_3 is usually described in terms of S=O double bonds. The bond lengths in the two oxides are almost the same when measured in the gas phase, and at 143 pm are among the shortest sulphur–oxygen bonds (Table 20). It seems at least true to say that the bond is closer to double than single.

Table 20 Bond lengths in sulphur–oxygen compounds

Species	Bond length/pm
SO	149
S_2O	146
SO_2	143
SO_3	143
asbestos-like $(SO_3)_3$	141, 161
SO_4^{2-} (in $CuSO_4$)	151
H_2SO_4	143, 153
$S_2O_4^{2-}$	151
$S_2O_6^{2-}$	145

SAQ 39 VSEPR theory predicts the shapes of the SO_2 and SO_3 molecules shown in Structures **43** and **44**, respectively. For each of these molecules, what spectroscopic method can be used to evaluate the bond distances shown?

43

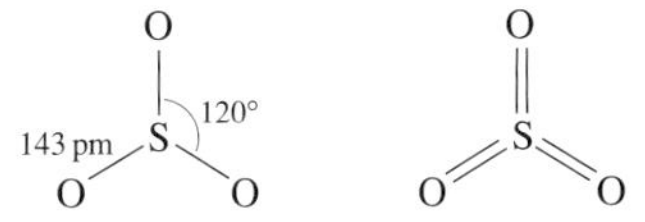

44

SULPHUR DIOXIDE

SO_2 is well known as the product of combustion of sulphur or sulphides in air or oxygen. It dissolves in water to give an acid solution

$$H_2O(l) + SO_2(g) = H^+(aq) + HSO_3^-(aq) \qquad (117)$$

much as carbon dioxide does, but there is no evidence for the existence of H_2SO_3 ('sulphurous acid') as a molecular species. The equilibrium position in Reaction 117 lies on the left, so SO_2 is evolved when acid is added to sulphites or hydrogen sulphites.

The standard redox potentials listed in Table 21 indicate the relative stability of the main oxidation numbers of sulphur in dilute acidic solutions.

Table 21 Redox reactions of some sulphur-containing species

Oxidation number	Sulphur species	Reaction	$E^{\ominus}$/V
+6	SO_4^{2-}	$SO_4^{2-}(aq) + 8H^+(aq) + 6e^- = S(s) + 4H_2O(l)$	+0.352
+4	SO_2	$SO_2(g) + 4H^+(aq) + 4e^- = S(s) + 2H_2O(l)$	+0.451
−2	H_2S	$S(s) + 2H^+(aq) + 2e^- = H_2S(g)$	+0.174

SAQ 40 (a) Subtract the second equation in Table 21 from the first to obtain the value of $E^{\ominus}$, and the equation, for the couple $SO_4^{2-}|SO_2$.

(b) Given $E^{\ominus}$ for the reduction of oxygen gas in acid solution:

$$\tfrac{1}{2}O_2(g) + 2H^+(aq) + 2e = H_2O(l); \qquad E^{\ominus} = 1.23\ V \qquad (118)$$

would atmospheric oxygen be thermodynamically capable of oxidizing SO_2 to SO_4^{2-}?

As you can see, sulphur in SO_2 is easily oxidized to oxidation number +6. Sulphur dioxide is consequently a useful reducing agent. The reaction between SO_2 and atmospheric oxygen is slow in the absence of a catalyst. Most of the undesirable effects of SO_2 pollution arise from catalysis of this oxidation (to produce H_2SO_4) by, for example, oxides of nitrogen from vehicle exhausts, or by particulate matter in polluted urban air. Case Study 4 investigates this in more detail.

The mild reducing properties of sulphur dioxide are put to good use commercially in bleaching natural fibres (straw, wool and newsprint). However, SO_2 is also capable of acting as an oxidizing agent in certain circumstances, such as the removal of H_2S from petroleum products (*Claus process*).

SAQ 41 Use Table 21 to predict the consequences of mixing H_2S and SO_2.

When condensed (b.t. −10 °C), SO_2 forms an interesting water-like solvent. Although it dissolves some ionic compounds fairly well, it is a better solvent for covalent organic compounds, particularly aromatic compounds*. This affinity for aromatic hydrocarbons is put to commerical use in the refining of lubricating oil, where liquid SO_2 is used to wash out traces of aromatic compounds.

Sulphur dioxide has long been used as an antiseptic and antioxidant in the food industry; wine casks have been fumigated with SO_2 for thousands of years. (The modern home winemaker who adds a sodium metabisulphite tablet to the wash water is following a very ancient practice! The sodium metabisulphite dissolves to release SO_2, which sterilizes the equipment.) It is still preferred to more potent modern competitors, because centuries of use have revealed no short- or long-term hazards in this form of food treatment.

SULPHUR TRIOXIDE

The reaction of SO_2 with oxygen in the gas phase:

$$SO_2(g) + \tfrac{1}{2}O_2(g) = SO_3(g); \qquad \Delta G^{\ominus}_m = -71.0\,\text{kJ mol}^{-1} \qquad (119)$$

or in aqueous solution, is thermodynamically favoured but kinetically hindered. From one point of view, the unfavourable kinetics are a good thing; SO_2 oxidation is not a reaction that we would wish to promote in the air of our cities! On the other hand, Reaction 119 is vitally important to chemical industry. Sulphuric acid, which is the product of the reaction of SO_3 with water, is the most heavily used industrial chemical (see Section 6.3.5).

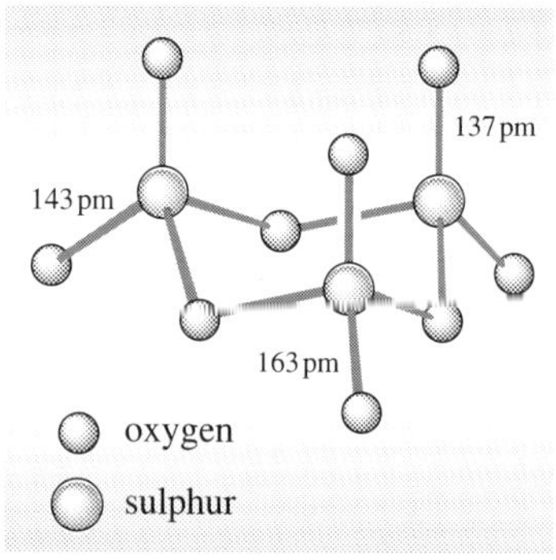

Figure 87 The ice-like form of the solid trimer of SO_3.

Sulphur trioxide is monomeric in the gas phase. However, at room temperature it exists as a white solid, which can occur in several polymeric forms. There is a low-melting trimer, S_3O_9 (Figure 87), which adopts the ice structure, as well as an involatile asbestos-like (Section 4.2.2) structure, in which chains of SO_4 tetrahedra are linked through apical oxygens. In this form, the bond lengths *in the chain* are close to single-bond distances (Table 20), whereas the terminal sulphur–oxygen bonds are double bonds, slightly shorter than those in SO_2.

As well as being a strongly acidic oxide, sulphur trioxide is also quite a strong oxidizing agent. It is used directly in the detergent industry as a sulphonating agent. The SO_3^- group thus introduced, confers water solubility on the long-chain hydrocarbon detergent molecule (see Structure **30**, p. 83).

6.3.5 OXOACIDS OF SULPHUR

We considered oxoacids in some detail in Section 5.6. This Section will now give us the opportunity to revise those rules and apply them to the sulphur oxoacids. With sulphur, remember, we have the added complication that an oxygen atom can be substituted by sulphur, giving a range of **thioacids**, for example thiosulphuric acid (Structure **46**), dithionic acid (Structure **47**) and polythionic acid (Structure **48**).

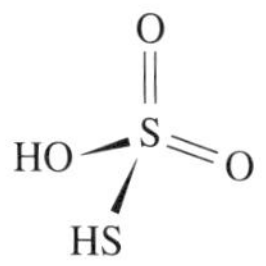

46

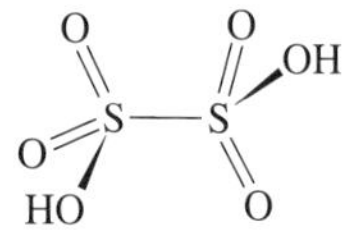

47

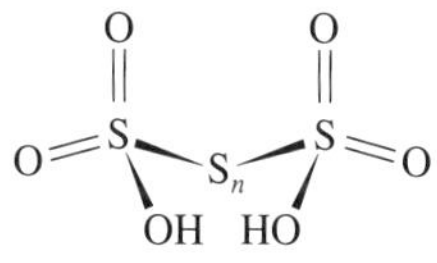

48

Table 22 shows the rich variety of oxoanions of sulphur which are possible. In fact, few of these oxoacids are known in the undissociated state. Apart from some of the polythionic acids, no oxoacid in which sulphur has an oxidation number <6 is stable, even in concentrated aqueous solution, and only sulphuric acid and its 'dimer' disulphuric acid exist as free acids at room temperature. The derivative obtained by replacing one of the —OH groups of sulphuric acid by an —SH group, thiosulphuric acid, is stable only below 0 °C.

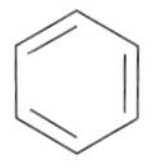

45

* An aromatic compound is a compound that contains a benzene (Structure **45**) or benzene-like ring.

Table 22 Oxoacids and oxoanions of sulphur*

Oxidation number	Oxoacid/oxoanion		
+6	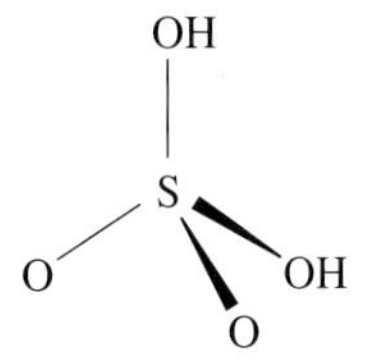 sulphuric	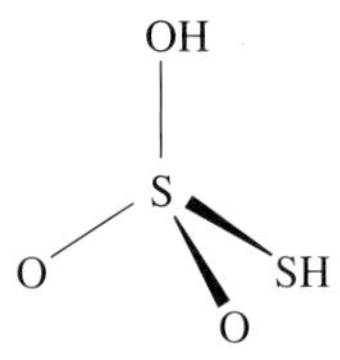 thiosulphuric	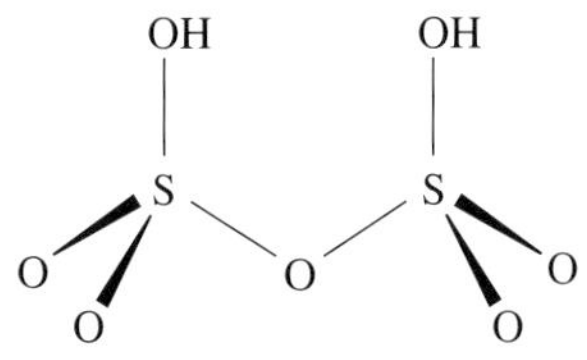 disulphuric
+5	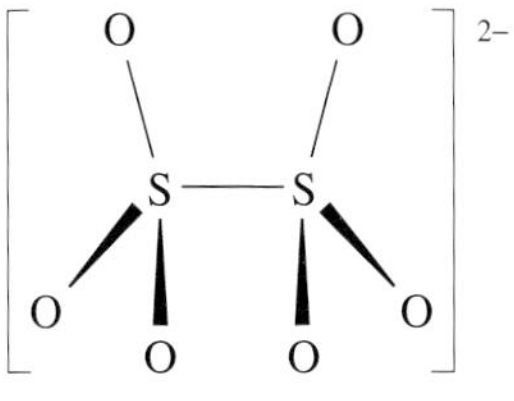 dithionate	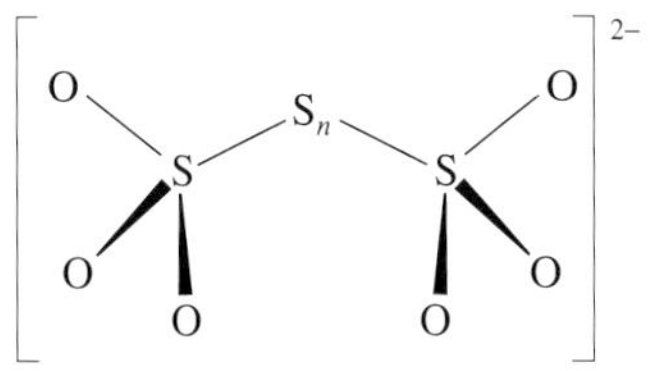 thionate (tri-, tetra-, penta-, hexa-)	
+4	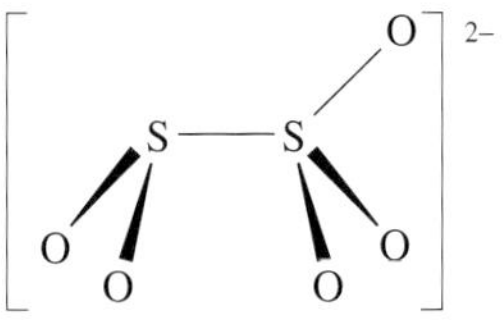 disulphite	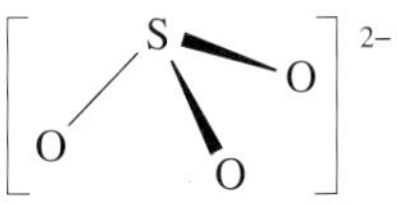 sulphite	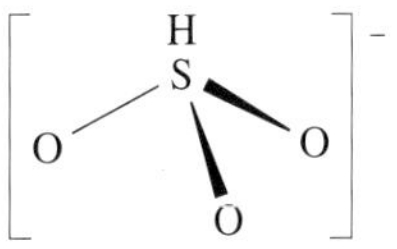 hydrogen sulphite
+3	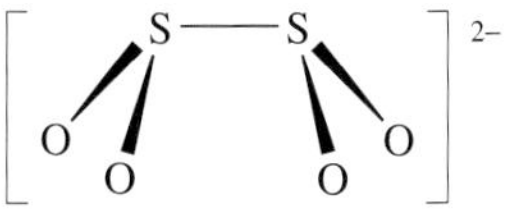 dithionite		

* These structures are shown to illustrate the versatility of bonding in sulphur oxoacids; there is no need to commit them to memory.

The nineteenth century claim that there is 'no better barometer to show the state of an industrial nation than the consumption of sulphuric acid per head of population' still holds true today, although the average person might be alarmed to learn that they 'consume' more than 20 times as much concentrated sulphuric acid as beer each year! (Interestingly, the per capita H_2SO_4 production costs are about the same as that of beer.)

The commercial process favoured at the present time, known as the **contact process**, uses either platinum or vanadium pentoxide, V_2O_5, as catalyst. Platinum is more effective, and allows a lower operating temperature, but is, of course, very expensive and is easily 'poisoned' by impurities. The SO_3 produced is dissolved in water, though not directly because of the violence of the reaction; it is first absorbed in concentrated H_2SO_4 to form disulphuric acid (oleum), $H_2S_2O_7$, which is then carefully diluted under controlled conditions to yield sulphuric acid:

$$H_2SO_4(l) + SO_3(g) = H_2S_2O_7(l) \qquad (120)$$

$$H_2S_2O_7(l) + H_2O(l) = 2H_2SO_4(l) \qquad (121)$$

Sulphuric acid is a very strong acid. Pure sulphuric acid is a colourless oily liquid, boiling at 317 °C, which, because of its high affinity for water, can dehydrate many organic compounds (recall the similar property of phosphoric oxide). Carbohydrates, for example, can be degraded to elemental carbon:

$$H_2SO_4(l) + C_6H_{12}O_6(s) = 6C(s) + H_2SO_4.6H_2O(l) \qquad (122)$$

Most of the salts of sulphuric acid (sulphates and hydrogen sulphates) are water soluble, with the notable exceptions of $SrSO_4$, $BaSO_4$ and $PbSO_4$; $CaSO_4$ is slightly soluble. The use of aluminium sulphate in the water industry was discussed in Section 3.2.2.

6.3.6 SULPHUR–CARBON AND SULPHUR–NITROGEN COMPOUNDS

A vast number of compounds is known in which sulphur forms single bonds with carbon of types such as RSH, RSR or RSSR, where R is an organic group. The more volatile of these compounds have powerful smells: hence the use of butanethiol, C_4H_9SH, to give an odour to natural gas, so that leaks are easily noticed.

As we noted earlier (Section 2.2), stable compounds with carbon–sulphur multiple bonds are somewhat more common than compounds with either carbon–silicon or carbon–phosphorus double bonds. The reaction of sulphur vapour with carbon at high temperatures gives carbon disulphide, S=C=S, which is isostructural with CO_2 (recall the discussion of the vibrational modes of CS_2 in Block 6); it is a smelly liquid at room temperature and a valuable solvent. The action of an electric discharge on CS_2 gives CS gas, which, unlike CO, polymerizes very quickly. Compounds of type $R_2C{=}S$, equivalent to ketones like propanone, $(CH_3)_2C{=}O$, are readily made, but many are unstable and polymerize to cyclic trimers with carbon–sulphur single bonds.

There is a very extensive chemistry of sulphur–nitrogen compounds involving both single-bonding and multiple-bonding between the elements. The most easily prepared compound is tetrasulphur tetranitride, S_4N_4, which can be made by passing NH_3 into a warm solution of S_2Cl_2 in benzene.

$$6S_2Cl_2 + 16NH_3 = S_4N_4 + 8S + 12NH_4Cl \qquad (123)$$

The crystals formed are yellow when cold, becoming orange at room temperature and red when heated: they are kinetically stable in air, but can decompose explosively to the elements if struck or heated. The structure of S_4N_4 is shown in Figure 88a. The S—N bond lengths are all the same, but quite short compared with the sum of the covalent radii (162 pm compared with 178 pm). It is difficult to describe the bonding in classical terms, but some possible resonance structures are shown in Figure 88b.

If S_4N_4 is heated and passed over silver wool at low pressure, an unstable dimer, S_2N_2, is formed, which has square-planar geometry (Structure **49**). The instability of S_2N_2 is demonstrated by striking or warming, but if left at room temperature it spontaneously polymerizes to give $(SN)_x$, a bronze-coloured compound with a metallic lustre, which

S
N
89.6°
90.4°
N
S

49

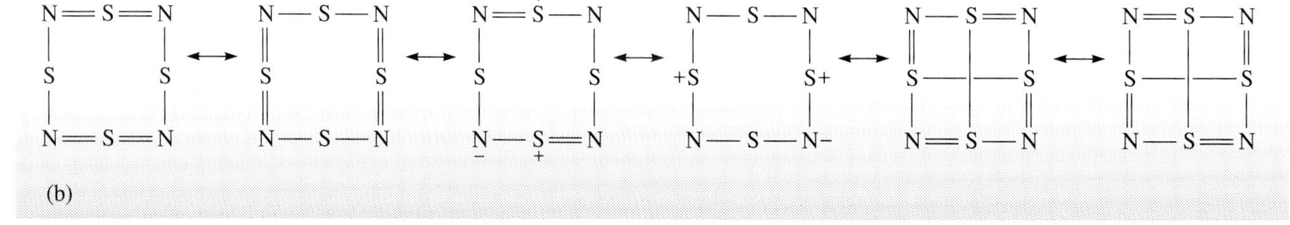

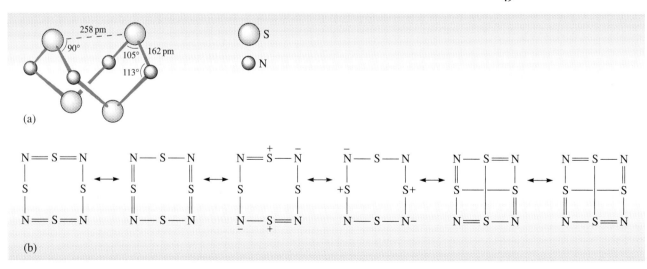

Figure 88 (a) The structure of S_4N_4; (b) possible resonance structures for S_4N_4.

contains almost planar S—N chains (Figure 89). Interest in these sulphur–nitrogen compounds was greatly fuelled by the discovery in the 1970s that polymeric $(SN)_x$ shows metallic conductivity down to liquid helium temperatures and becomes a superconductor when cooled below 0.3 K. Although this is a very low temperature for the onset of superconductivity, it was a very exciting discovery at the time, because this was the first superconductor found that did not contain a metal. The electrical conductivity is much greater along the direction of the chains; compounds such as this are known as *one-dimensional metals*.

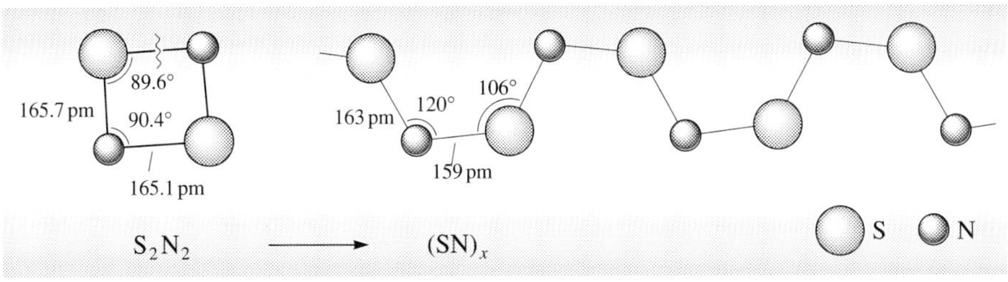

Figure 89 The structure of $(SN)_x$ and its relation to S_2N_2.

6.4 SELENIUM, TELLURIUM AND POLONIUM

We have already noted that the properties of the elements in Group VI show the characteristic trends that we have come to expect in descending a Group. The elements become more metallic in character: oxygen is a covalently bonded gaseous diatomic molecule; sulphur is a solid containing S_8 molecules and is an insulator; selenium (non-metal) and tellurium (semi-metal) are semiconductors with polymeric structures; polonium is a metal. The compounds of selenium, tellurium and polonium also illustrate the inert pair effect and the tendency to higher coordination numbers.

In recent years selenium has become very important in the photocopying industry, which makes use of its ability to act as a photoconductor. Incident light is able to excite electrons across its small band gap (1.8 eV for amorphous selenium). In a photocopying machine, a film of amorphous selenium is deposited on the copier plate and this is given a positive charge electrostatically. During the copying process, the plate is exposed to a light and dark pattern, and in the light areas the plate becomes discharged due to the photoelectrons produced. To develop the image, black toner particles are used which are attracted to the charged areas on the plate, and which correspond to the dark areas of the original pattern. The toner can then be transferred to paper and fixed, usually by heating, to make the final copy. The main uses of elemental selenium, however, lie in the glass industry, where elemental selenium is used as a decolorizer of glass, and in the mixed sulphide/selenide of cadmium, Cd(S/Se), which gives the best ruby red coloration to glass.

Tellurium is not very important industrially, and is mainly used in the production of steel and other alloys.

Moving down Group VI, we see, as we have seen for Groups IV and V, a decreasing ability to form double bonds. Take the compounds with carbon as an example. Carbon dioxide is a gas at room temperature, which exists only in the form of O=C=O molecules. In contrast, carbon disulphide and carbon diselenide are liquids at room temperature, which contain S=C=S and Se=C=Se molecules. However, under high pressure, and on heating, CS_2 gives a black solid polymer containing both double and single bonds; with CSe_2, this change occurs slowly at room temperature and pressure; Te=C=Te is unknown.

The halides formed by selenium, tellurium and polonium are summarized in Table 23. It should be remembered that polonium is a rather rare element, which is difficult to handle because of its radioactivity; because of this its chemistry has not been as fully explored as that of selenium and tellurium.

Table 23 The halides of selenium, tellurium and polonium

Oxidation number	<1	+1	+2	+4	+5	+6
		Se_2F_2	SeF_2*	SeF_4		SeF_6
		Se_2Cl_2	$SeCl_2$*	$(SeCl_4)_4$		
		Se_2Br_2	$SeBr_2$*	$(SeBr_4)_4$		
				TeF_4		TeF_6
	Te_2Cl		$TeCl_2$	$(TeCl_4)_4$		
	Te_3Cl_2					
	Te_2Br		$TeBr_2$*	$(TeBr_4)_4$		
	Te_2I	Te_4I_4		$(TeI_4)_4$		
				PoF_4?		
			$PoCl_2$	$PoCl_4$		
			$PoBr_2$	$PoBr_4$		
			PoI_2	PoI_4		

* Unstable at room temperature.

Table 23 shows evidence for the trends that we have come to expect. The highest oxidation number (+6) is stabilized by fluorine only in the cases of selenium and tellurium; PoF_6 is not known. There are no low oxidation number fluorides apart from F—Se—Se—F and Se=SeF_2, and these can only be trapped at low temperatures. Selenium forms no iodides, but the more electropositive tellurium and polonium do.

We see the increasing inert pair effect down the Group as the increased stability of oxidation number +4. All four tetrahalides exist for tellurium and polonium, but not for selenium.

The tendency towards higher coordination number down the Group can be seen in SeF_4 and TeF_4. Thus, SeF_4 is a colourless volatile liquid, which gives monomeric four-coordinate $\mathbf{C}_{2v}$ molecules in the gas phase (like SF_4), whereas TeF_4 forms colourless crystals containing chains of *cis*-linked five-coordinate square-pyramidal TeF_5 units (Figure 90). Both compounds are highly reactive, fuming in moist air to give HF. (PoF_4 is not well characterized.)

The oxides in which the element is in oxidation number +6, MO_3, are known for Se and Te. The structure of SeO_3 contains cyclic tetramers in which the selenium is four coordinate, and the α-TeO_3 structure contains TeO_6 octahedra sharing all vertices to give a three-dimensional lattice; again we see the trend to higher coordination number in descending the Group. All three dioxides, MO_2, are known; use SAQ 47 to try to predict the changes in structure that you might expect to see as you go down Group VI.

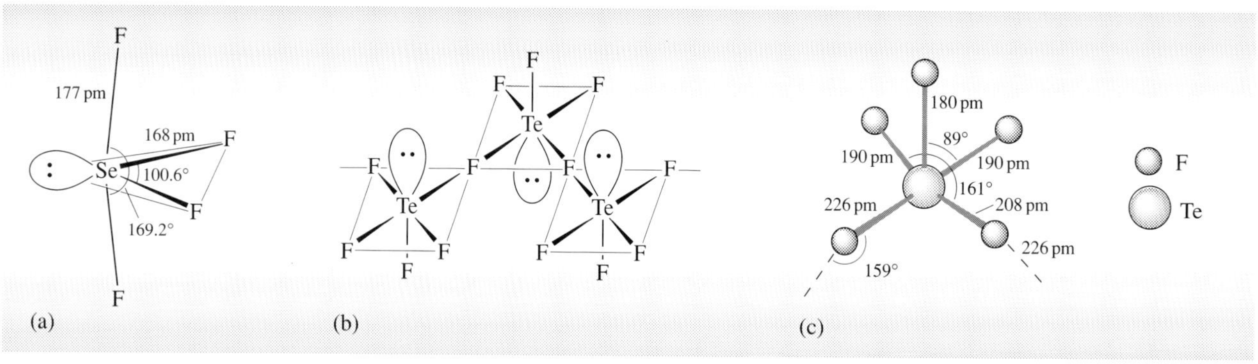

Figure 90 The structures of: (a) gaseous SeF_4; (b) solid TeF_4; (c) geometry around a TeF_5 group in solid TeF_4.

6.5 SUMMARY OF SECTIONS 6.3 AND 6.4

1 Sulphur forms hydrides, called sulphanes, only in oxidation number +2 (or lower). These have the general formula H_2S_n (where n represents the number of sulphur atoms in the chain).

2 The higher sulphanes are unstable with respect to decomposition into H_2S and solid sulphur.

3 The sulphanes are weak acids, liberated by acidification of their salts (sulphides or polysulphides).

4 Sulphur can occur in many oxidation numbers in its halogen compounds. The highest oxidation number achieved is graded in order of electronegativity (or oxidizing power) of the halogen: for bromosulphanes it is +2, for chlorosulphanes +4, for fluorosulphanes +6 (there are no known iodosulphanes).

5 Halosulphanes (apart from SF_6, which is kinetically inert) are readily hydrolysed, and fume in moist air due to formation of HX.

6 The important oxides of sulphur are the dioxide, SO_2, and trioxide, SO_3. As with fluorine, use of an electronegative ligand favours higher oxidation numbers.

7 SO_2 and SO_3 are both acidic oxides; SO_3 dissolves in water with considerable evolution of heat to form sulphuric acid, H_2SO_4.

8 SO_2 is frequently used as a mild reducing agent (in the textile, food and paper industries), but it also has oxidizing properties, which are made use of in the Claus process for removal of H_2S from petroleum products.

9 SO_2 is catalytically oxidized by $O_2(g)$ to SO_3 in the contact process for sulphuric acid manufacture.

10 Polymeric $(SN)_x$ is a metallic compound, which conducts electricity along the chains (one-dimensional) and becomes superconducting below 0.3 K.

11 Sulphuric acid accounts for around nine-tenths of all the sulphur used industrially, and most of it is used to make phosphate and ammonium sulphate fertilizers.

12 The chemistry of selenium, tellurium and polonium illustrate trends down a Group such as increasing metallic character, the inert pair effect and the tendency to higher coordination numbers.

SAQ 42 Predict the formula of the oxoacid of sulphur in which sulphur is in its highest oxidation number. What is the formula of the condensed acid made from two molecules of this acid?

SAQ 43 Using Pauling's rules, predict the strength of sulphuric acid, H_2SO_4. If its first dissociation constant is $1 \times 10^3\,\text{mol l}^{-1}$, what value would you predict for the second dissociation constant?

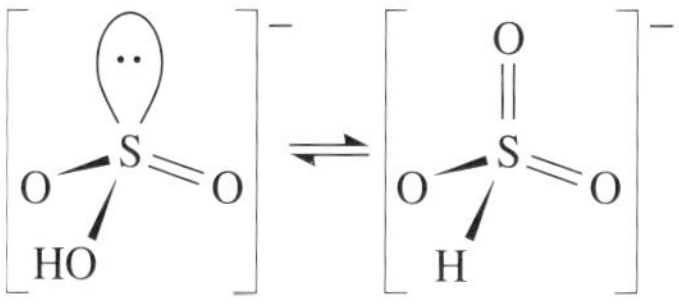

50 **51**

SAQ 44 It has been claimed that in aqueous solution the hydrogen sulphite ion, HSO_3^-, consists of two forms, Structures **50** and **51**, in equilibrium. Does the Raman spectrum of $HSO_3^-(aq)$ (Figure 91) provide any evidence in support of this claim?

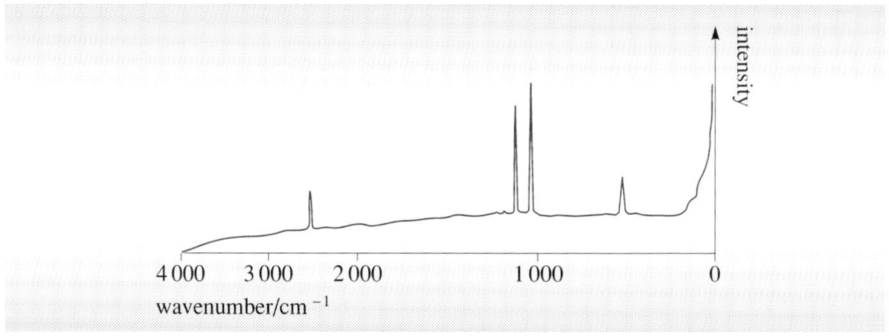

Figure 91 The Raman spectrum of $HSO_3^-(aq)$.

SAQ 45 S_4N_4 decomposes explosively. Suggest the likely decomposition route, and therefore what drives this reaction.

SAQ 46 Suggest some resonance structures to describe the bonding for the S_2N_2 molecule.

SAQ 47 How might you expect the structure of the Group VI oxides, MO_2, to vary down the Group?

SAQ 48 NO is a well-known gas, but NS has no independent existence. Why?

7 THE TYPICAL ELEMENTS: TRENDS IN THE PERIODIC TABLE

You have now completed your study of the descriptive chemistry of the typical elements, which are laid out in the mini-Periodic Table of Figure 92.

The descriptive chemistry of Groups I and II of this table was covered in Block 3, and that of Groups VII and 0 in Block 5. This Block has completed the job by dealing with Groups III–VI. As this information has accumulated in Blocks 3, 5 and 7, certain trends in chemical properties that occur both across the rows, and down the Groups of Figure 92 have emerged. In this last Section we summarize and remind you of some of these trends.

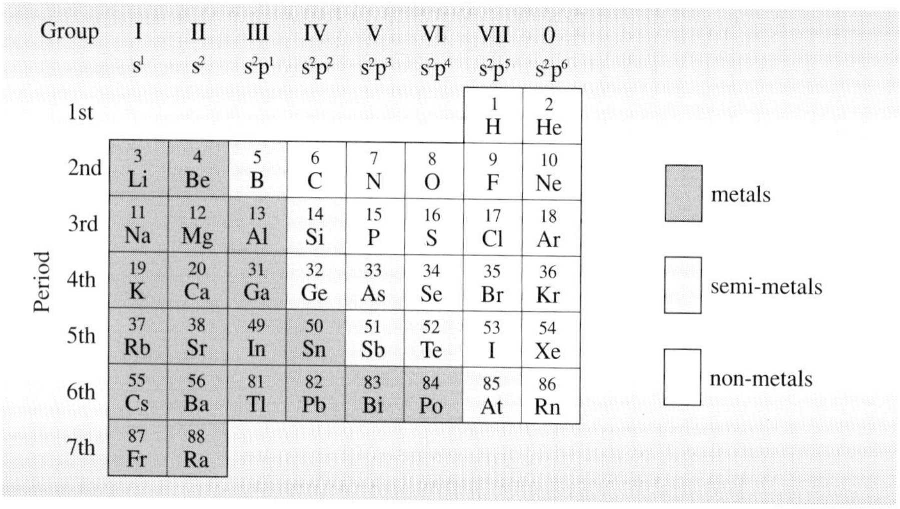

Figure 92 A Periodic arrangement of the typical or main-Group elements (note that the helium atom has the outer electronic configuration s^2 rather than s^2p^6).

7.1 TRENDS ACROSS THE ROWS

7.1.1 IONIZATION ENERGIES, ELECTRONEGATIVITIES AND COVALENT RADII

In Block 3, Sections 3 and 4, we showed that the first ionization energies and electronegativities of the typical elements showed overall increases across the rows of Figure 92. By contrast, the single-bond covalent radii showed a decrease. All three trends reveal the importance of the increase in the nuclear charge as successive protons are added to the nucleus of the atoms. Such an increase causes the outer electrons to be bound more and more tightly across the row. All these tendencies are summarized in more detail on p. 26 of Block 3. They are trends in the properties of atoms, and as such provide possible explanations of other trends in physical and chemical properties. To these we can now turn.

7.1.2 METALS, SEMI-METALS AND NON-METALS

At the beginning of a row of Figure 92, the nuclear charge and ionization energies are relatively low. Electrons are then more easily lost to, or become part of, an 'electron pool' or 'electron gas', which, according to the simplest theory of metallic bonding, holds the metal together through its interaction with the residual positive ion cores (Figure 93). The increase in nuclear charge therefore accounts for the tendency for metals to give way first to semi-metals, and then to non-metals, across a row of Figure 92 (Block 3, Sections 4.1 and 4.2). This tendency is illustrated, for example, by the properties of the elements of Period 3, some of whose structures are shown in Figure 94.

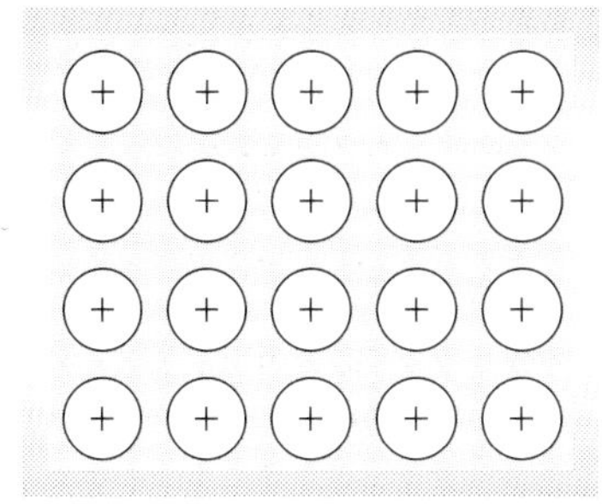

Figure 93 In the simplest theory of metallic bonding, the grey electron gas acts as a kind of glue, which holds the metal together through its interaction with the positive ion cores.

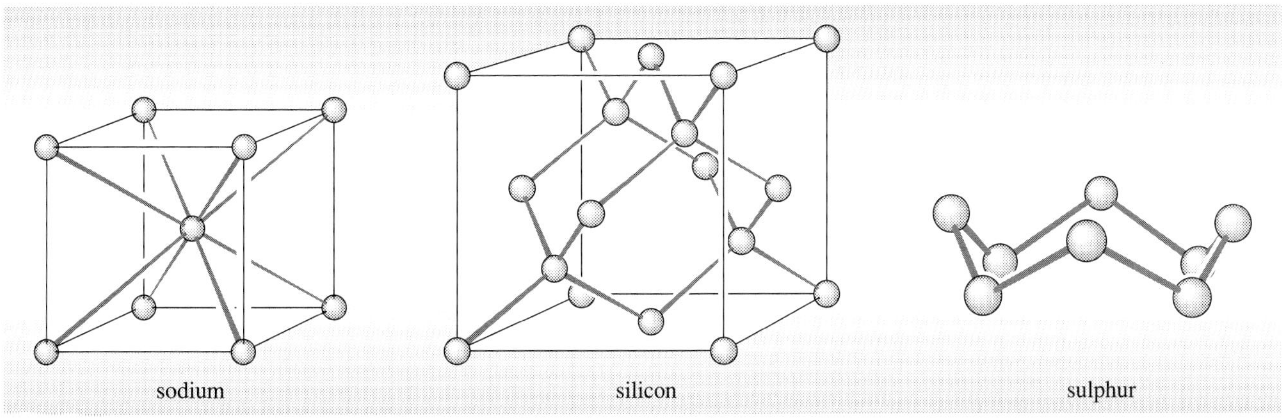

Figure 94 Sodium, silicon and sulphur are in the same row of the Periodic Table. The approximate electrical conductivities of sodium, silicon and sulphur at 0 °C are 10^7 S m^{-1}, 10 S m^{-1} and 10^{-15} S m^{-1}, respectively. Consistent with its high metallic conductivity and other metallic properties, sodium has a characteristic body-centred metallic structure. According to the simplest theory, this consists of Na^+ cores and an electron gas. At first sight the increase in nuclear charge at silicon has confined the valence electrons to two-electron covalent bonds in a diamond-type structure. However, as you saw in Case Study 2, the confinement is not so great that thermal energies at normal temperatures cannot detach *some* of these bonding electrons and transfer them to an electron gas (simple theory) or to a conduction band (band theory). Silicon is therefore a semi-metal and a semiconductor. By contrast, at sulphur, confinement of the valence electrons to S—S covalent bonds or lone pairs is complete, and the structure consists of discrete S_8 ring molecules, between which van der Waals forces operate.

7.1.3 THE STRUCTURES OF HALIDES, HYDRIDES AND OXIDES

At the beginning of a row in Figure 92, the electronegativity of an element is low, much lower than that of a halogen, oxygen or hydrogen. Halides, oxides and hydrides then form ionic structures; thus, the halides and hydrides of lithium, sodium, potassium and rubidium all crystallize with the structure of NaCl, and the oxides of these elements have the antifluorite structure (Figure 95).

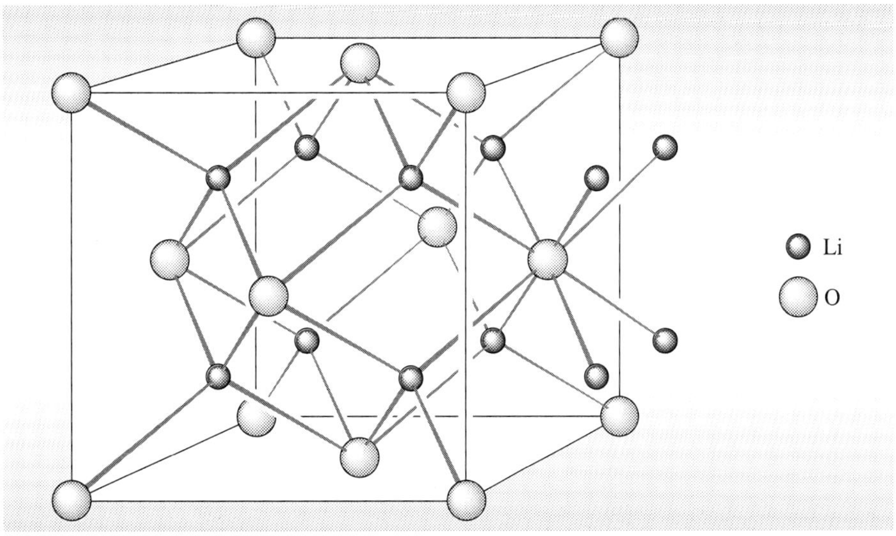

Figure 95 The antifluorite structure of the oxides Li_2O, Na_2O, K_2O and Rb_2O, shown here for the case of Li_2O. It is what would be expected of a collection of ions: each lithium is surrounded by four oxygens, and each oxygen by eight lithiums.

In moving across the rows, the electronegativity of the elements increases, and there is a corresponding decrease in the electronegativity difference between the two kinds of atom in their halides and oxides. Consequently, the ionic character of the compounds decreases. Structurally, this is apparent in the way that three-dimensional ionic structures give way first to either layer, chain or macromolecular structures, and then to discrete covalent molecules. This tendency has already been touched on for hydrides in Block 5, Section 6.1, and for halides in Block 3, Section 13.3. To take the specific example of chlorides in the third Period, sodium chloride, which is typically ionic, gives way to $MgCl_2$, which adopts a $CdCl_2$ layer structure (Figure 96), and then to $AlCl_3$ with the layer structure described in Figure 25 of this Block.

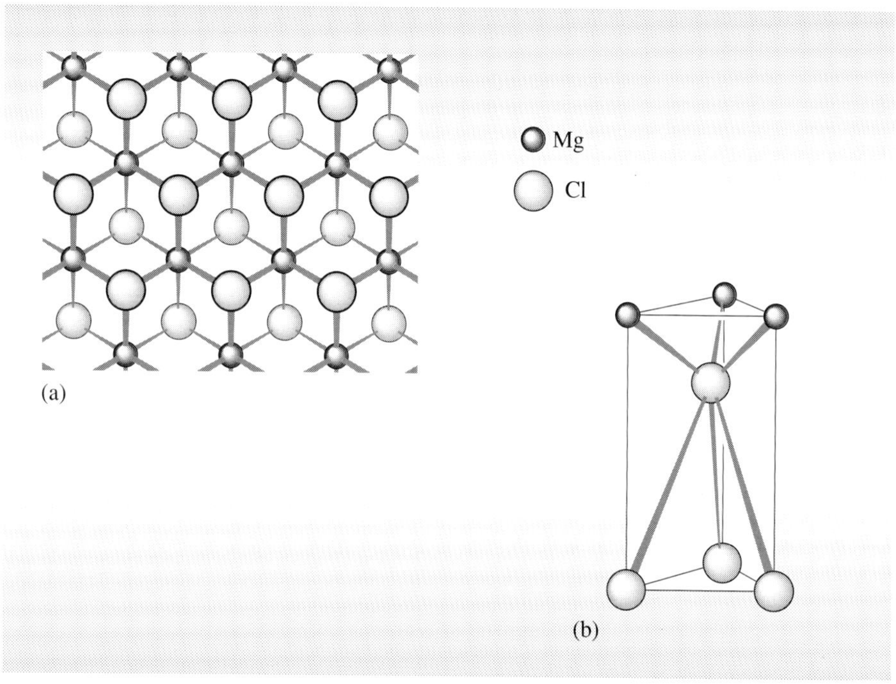

Figure 96 The $CdCl_2$ layer structure of $MgCl_2$. In (a) one layer is viewed from above; it has three decks. The top and bottom decks consist of chlorines; these have been distinguished in the Figure by giving the top-deck atoms a heavy border; the middle deck consists of coloured, octahedrally coordinated magnesiums. In (b) the environment of each chlorine is shown. It is a distorted trigonal prism: there are three magnesiums on one side, but on the other side there are three chlorines substantially further away in an adjacent layer. This is not what would be expected of a collection of ions.

The loss of ionic character between NaCl and $AlCl_3$ is also apparent in the conductivity of the molten salts: molten NaCl and $MgCl_2$ are excellent conductors of electricity; molten $AlCl_3$ is a very poor conductor. Beyond aluminium, there are chlorides with discrete covalent molecules. Thus, $SiCl_4$ and SCl_2 are non-conducting volatile liquids containing tetrahedral and V-shaped molecules, respectively, and Cl_2 is a gas.

■ How do trends of this kind reveal themselves in the structures of the normal oxides of sodium, silicon and chlorine?

□ Na_2O has an ionic-type antifluorite structure (Figure 95), SiO_2 has a macromolecular structure (Figures 6 and 43) in which silicon and oxygen attain noble gas structures by forming covalent bonds with each other, and the oxides of chlorine (Cl_2O, ClO_2, Cl_2O_7) consist of discrete covalent molecules (Block 5, Figure 61).

7.1.4 TRENDS IN THE FORMULAE OF OXIDES AND HYDRIDES

It is worth recalling the trends in the formulae of the highest normal oxides and of the highest hydrides, which were so successfully exploited by Mendeléev when he put together his Periodic Table. As we saw in Block 3, Section 2, the elements in Figure 92 usually exercise their highest valency and highest oxidation number in the highest normal oxide. In most cases, both are equal to the Group number and to the number of outer electrons, which is taken to be zero for the noble gases.

■ What are the formulae of the highest normal oxides of the elements of the third Period?

□ Na_2O, MgO, Al_2O_3, SiO_2, P_2O_5, SO_3, Cl_2O_7 and no oxide at argon; the highest valencies and highest oxidation numbers are equal to the Group numbers, and run from one at sodium to seven at chlorine.

If we allow expansion of the octet as in Section 2.2, this valency pattern occurs because in its highest oxide, each element uses all its outer electrons in forming bonds. Thus, with six outer electrons, sulphur can form six electron-pair covalent bonds, and an oxide SO_3 is formed (Figure 87).

In the case of the hydrides the pattern is different (Block 5, Section 6.1). There is no expansion of the octet, and the elements attain noble gas configurations in the hydrides. This gives formulae AH_x, where x takes values 1, 2, 3, 4, 3, 2, 1, 0 across a row.

7.1.5 ACID–BASE PROPERTIES OF OXIDES AND HYDRIDES

There is a tendency for normal oxides, especially the highest normal oxides, to change from basic, through amphoteric to acidic from left to right across a Period of Figure 92. This trend was discussed in Section 6.2.2 and is pictured in Table 18. Thus, at the start of the third Period, Na_2O is basic, and at the end, Cl_2O_7 is acidic. This is apparent from the reactions with water, when alkaline and acid solutions are formed, respectively:

$$Na_2O(s) + H_2O(l) = 2Na^+(aq) + 2OH^-(aq) \tag{124}$$

$$Cl_2O_7(l) + H_2O(l) = 2H^+(aq) + 2ClO_4^-(aq) \tag{125}$$

These reactions were considered in Block 3, Section 9.2 and Block 5, Section 9.1.2, respectively.

In between these two oxides comes Al_2O_3, which has negligible solubility in pure water, but reveals amphoteric character (Section 3.2.1) by dissolving in both acids and alkalis:

$$Al_2O_3(s) + 6H^+(aq) = 2Al^{3+}(aq) + 3H_2O(l) \tag{126}$$

$$Al_2O_3(s) + 2OH^-(aq) + 3H_2O(l) = 2[Al(OH)_4]^-(aq) \tag{127}$$

A related trend is apparent in the reactions of the hydrides with water. At the beginning of the Periods, the Group I and Group II hydrides give hydrogen gas and alkaline solutions with water (Block 5, Section 6.1); for example

$$NaH(s) + H_2O(l) = Na^+(aq) + OH^-(aq) + H_2(g) \tag{128}$$

These basic properties contrast with the acidity of the hydrides of Group VII elements; for example

$$HCl(g) + H_2O(l) = H_3O^+(aq) + Cl^-(aq) \tag{129}$$

7.2 TRENDS DOWN THE GROUPS

7.2.1 IONIZATION ENERGIES, ELECTRONEGATIVITIES AND SIZE

In Block 3, Sections 3 and 4, we showed that the first ionization energies and electronegativities of the typical elements tended to decrease down a Group. By contrast, measures of size such as single-bond covalent radii or the ionic radii of similarly charged ions increased. Figures 97 and 98 show the contrasted examples of ionization energy and ionic radii.

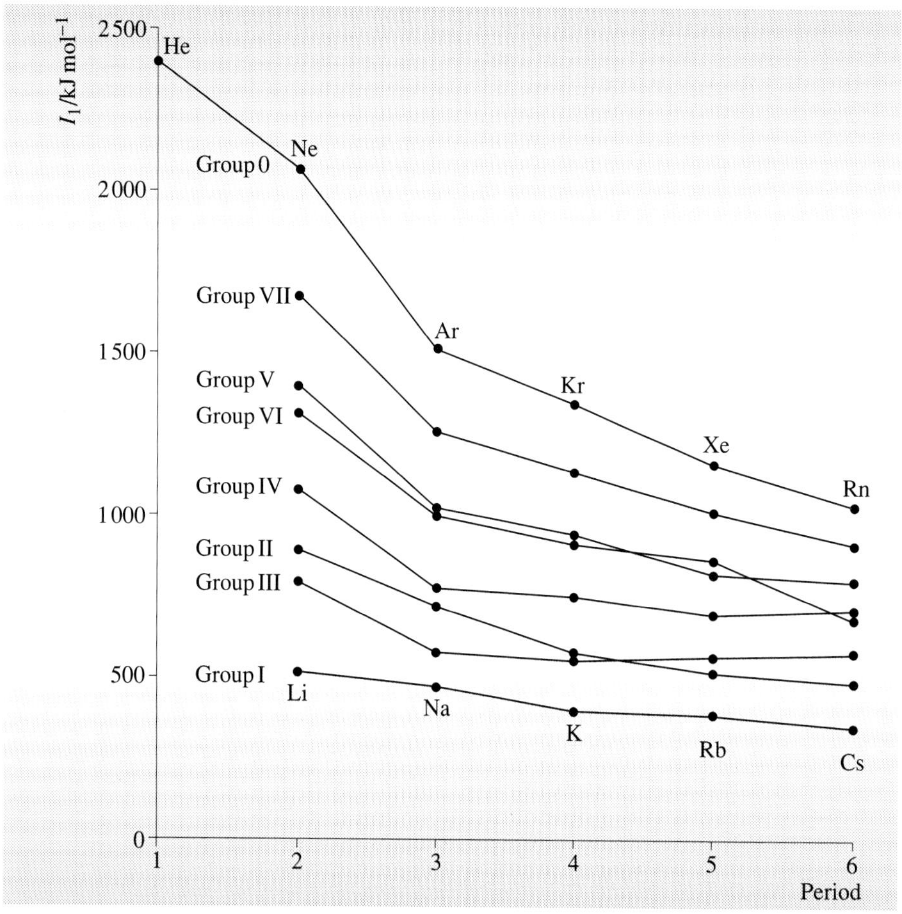

Figure 97 The first ionization energies of the typical elements tend to decrease down a Group.

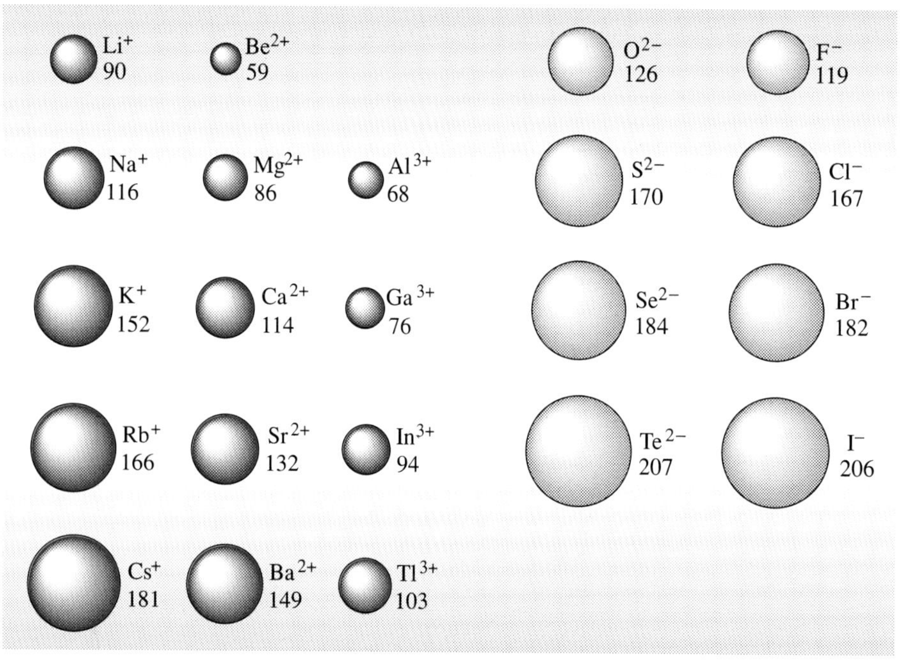

Figure 98 The ionic radii of similarly charged ions of the typical elements tend to increase down a Group.

All these trends reveal the importance of the increase in distance of the outer electrons from the nucleus as the successive electron shells build up. This increasing distance weakens the hold that the nucleus has on the outer electrons.

7.2.2 METALS, SEMI-METALS AND NON-METALS

At the top of the Groups, the outer electrons are closer to the nucleus, and ionization is difficult. Electrons are therefore less likely to be lost to become part of an electron pool that can engender metallic bonding as in Figure 93. The increasing distance of the outer electrons from the nucleus therefore accounts for the tendency for non-metals to give way first to semi-metals, and then to metals, down a Group of Figure 92. Group IV provides a good example: carbon is a non-metal, silicon and germanium are semi-metals and semiconductors, and tin and lead are metals.

7.2.3 STRUCTURE AND BONDING IN HALIDES AND OXIDES

Within the Groups in Figure 92, the electronegativity is highest at the top, and it is here where it is usually closest to the high electronegativities of oxygen and the halogens. On descending a Group, electronegativity decreases, so electronegativity differences in oxides and halides tend to increase. Oxides and halides should therefore become more ionic.

In the Group II chlorides this trend is marked first by a progression from the chain structure of $BeCl_2$ (Figure 99) to the layer structure of $MgCl_2$ (Figure 96). Neither are what we would expect of a collection of ions.

Dichloride structures more consistent with an ionic model are observed only for calcium, strontium, barium and radium: $CaCl_2$ has the rutile structure (Figure 100), $SrCl_2$, the fluorite structure, and, in $BaCl_2$ and $RaCl_2$, each metal atom is surrounded by nine chlorides (Figure 101), and each chloride by either four or five metal atoms.

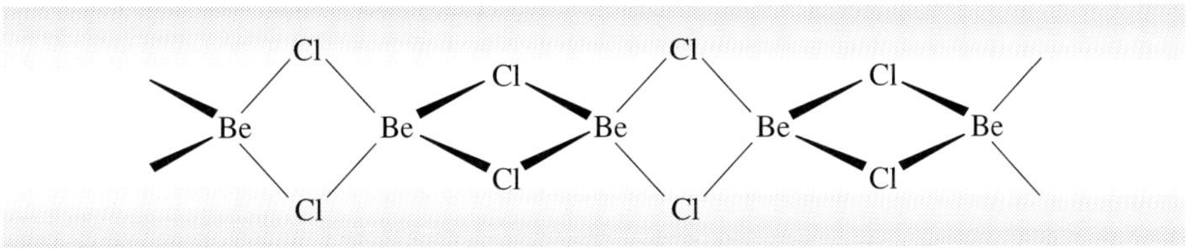

Figure 99 The chain structure of solid $BeCl_2$.

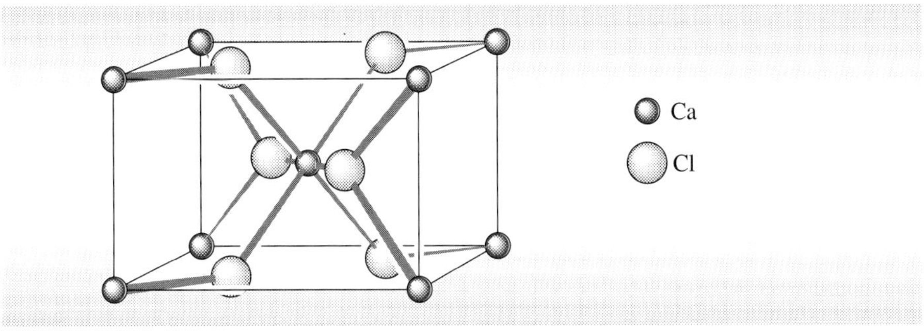

Figure 100 The rutile structure of $CaCl_2$.

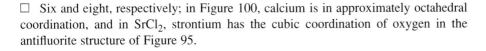

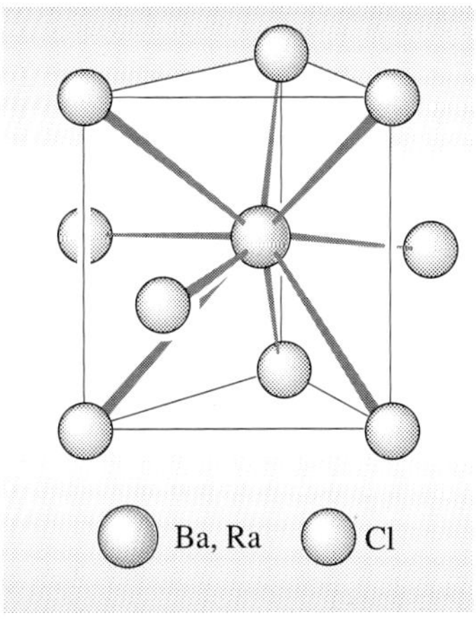

Figure 101 The ninefold coordination of barium and radium in $BaCl_2$ and $RaCl_2$. There are six chlorides at the corners of a surrounding trigonal prism, and another three out (capping) through the centres of the vertical rectangular faces.

■ What is the coordination number of calcium in $CaCl_2$, and of strontium in $SrCl_2$?

□ Six and eight, respectively; in Figure 100, calcium is in approximately octahedral coordination, and in $SrCl_2$, strontium has the cubic coordination of oxygen in the antifluorite structure of Figure 95.

Thus, along with the structural evidence of increasing ionic character, we see the influence of the increase in ionic radii down the Group: the coordination number increases from four in $BeCl_2$ to six in $MgCl_2$ and $CaCl_2$, through eight in $SrCl_2$ to nine in $BaCl_2$ and $RaCl_2$.

A similar trend in a very different form is revealed in the Group IV dioxides. Carbon dioxide is a gas consisting of linear covalent molecules, O=C=O, whereas SiO_2 and GeO_2 form quartz-like structures (Figure 6), consistent with the presence of Si—O and Ge—O covalent bonds; SnO_2 and PbO_2, however, have the rutile structure. Here, there is a progression from a discrete molecular covalent substance, through macromolecules to ionic structures. At the same time, the coordination number increases from two in CO_2, to four in SiO_2 and GeO_2, to six in SnO_2 and PbO_2. The progression includes, and is partly explained by, the important difference noted in Section 2.1: multiply bonded compounds are more common for second-row than for later-row elements. Because carbon forms double bonds in its dioxide, the coordination number is lower than in the singly bonded SiO_2 and GeO_2.

In the case of the Group VI dioxides, the trend is recognizably similar. It was examined in SAQ 47. There is a transition from covalent molecules at O_3 and SO_2, through macromolecular chain structures in SeO_2 and TeO_2 (Figure 102) to a fluorite, ionic-type structure in PoO_2. The coordination number increases in the sequence two, two, three, four, eight, but the drift away from π-bonding is in this case more gradual, and is completed only at tellurium.

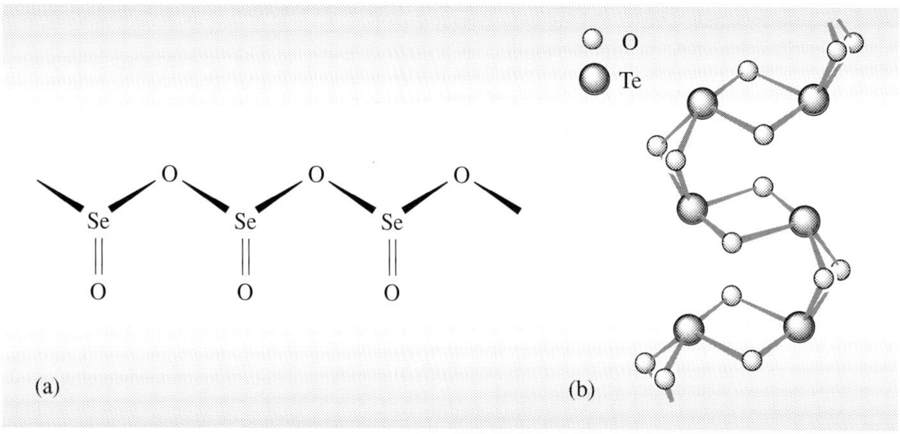

Figure 102 The macromolecular chain structures of (a) SeO_2 and (b) tellurite, the naturally occurring form of TeO_2. In both structures, the Group VI element has been drawn in a tetravalent state.

7.2.4 ACID–BASE PROPERTIES OF OXIDES

Normal oxides of a particular formula type tend to become less acidic, and more basic in descending a Group. This is apparent from Table 18. Thus, in Group II, BeO is amphoteric (Block 3, Section 13.2): it has very low solubility in water, but dissolves in both acids and alkalis. The other Group II monoxides, however, are basic, and yield alkaline solutions when added to water; for example

$$BaO(s) + H_2O(l) = Ba^{2+}(aq) + 2OH^-(aq) \qquad (130)$$

The full change from acidic to basic is best seen in Group III (Section 3.3). The oxide B_2O_3 is acidic, Al_2O_3 and Ga_2O_3 are amphoteric, and In_2O_3 and Tl_2O_3 are basic.

7.2.5 SPECIAL DIFFERENCES BETWEEN THE SECOND AND SUBSEQUENT PERIODS

In Figure 92, the second Period is the first to contain an element from each Group, and the differences between each of its elements, and the one beneath it in the Table, are greater than in any subsequent Period. We concentrate here on the particular differences mentioned in Sections 2.1 and 2.2. One of these was the greater preference for multiply bonded structures in the chemistry of second-row elements. This is especially obvious in the forms of the elements of Groups IV–VI. Thus, nitrogen and oxygen are gases, but phosphorus and sulphur assume solid, singly bonded forms at room temperature (Sections 5 and 6).

■ To what extent is this trend apparent in the known forms of the elements carbon and silicon?

□ See Figure 103 and its caption.

Although molecules containing silicon–silicon and phosphorus–phosphorus multiple bonds are known (Sections 4.3.4 and 5.5.6, respectively), they are thermodynamically unstable, and exist only when bulky groups attached to the silicon or phosphorus atoms provide a kinetic barrier to conversion into single-bonded polymers.

Two distinct factors tend to make π-bonded structures more stable in the second Period than the third. Firstly, the strength of π-bonding seems to decrease down a Group; secondly, in Groups V and VI, alternative single-bonded structures are disadvantaged by a weakness of the N—N and O—O single bonds, which is attributed to the repulsion between non-bonding electrons over the necessarily short internuclear distances (Block 5, Figure 48 and SAQ 19).

Another difference between second- and third-row elements was discussed in Section 2.2: third-row typical elements in Groups V–VII can exist in higher oxidation numbers. For example, the highest fluorides of oxygen and sulphur are OF_2 and SF_6; the highest normal oxides of fluorine and chlorine are F_2O and Cl_2O_7 (Block 5, Section 9.1). One possible explanation considered in Section 2.2 was the expansion of the octet in third and later Periods through the use of d orbitals of the same principal quantum number as the s and p valence electrons, a possibility that is not open to second-row elements. This may allow higher coordination numbers to be reached by second-row elements in both oxidation reactions and complex ion formation. Thus, potassium fluoride and silicon tetrafluoride will react to form K_2SiF_6, which contains the octahedral complex ion, $[SiF_6]^{2-}$. The silicon octet in SiF_4 has been expanded by the donation of lone pairs by two fluoride ions, giving silicon a coordination number of six. As this possibility is not open to carbon, CF_4 does not react with KF, and K_2CF_6 has not been prepared. In Section 4.3.1, the ability of silicon to expand its coordination number provided a possible reason why the hydrolysis of $SiCl_4$ is fast, but that of CCl_4 is immeasurably slow: as CCl_4 is less willing to allow entry of an incoming ligand into its coordination sphere, there is a kinetic barrier to the thermodynamically favourable hydrolysis.

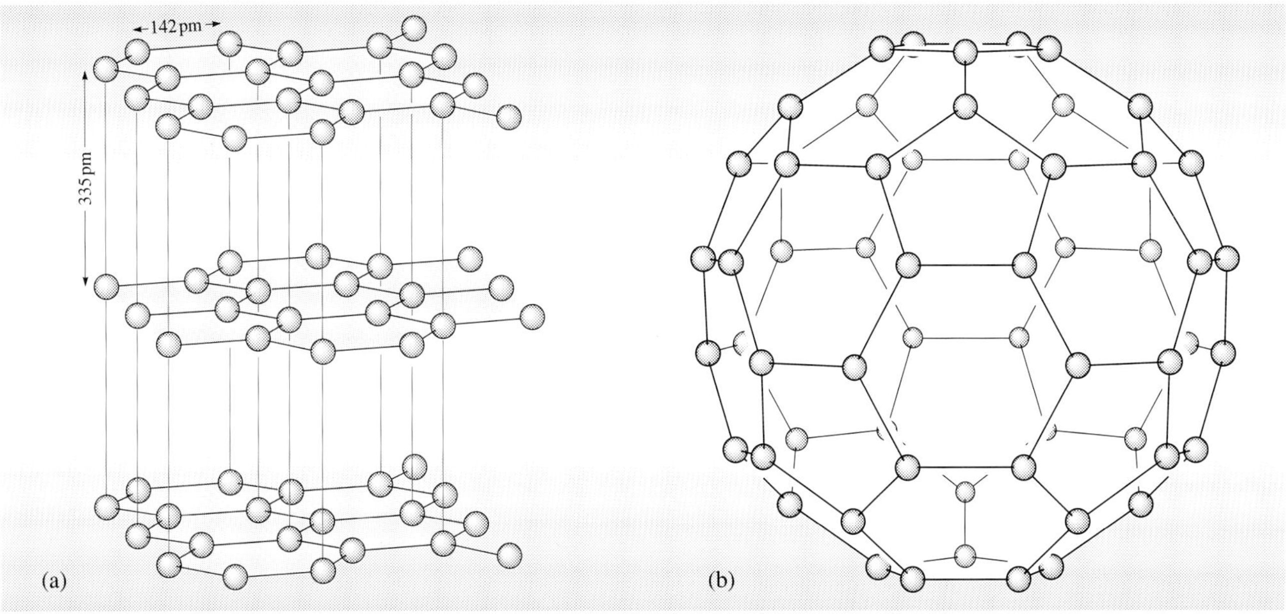

Figure 103 Carbon, in the form of diamond, and silicon share the same singly bonded structure, but carbon is also found as (a) graphite and (b) fullerenes, which consist of shared π-bonded rings.

7.2.6 THE INERT PAIR EFFECT

If one ignores catenated compounds (those in which atoms of the element in question are bound to each other), the *typical* elements tend to form oxidation numbers that differ by multiples of two. Thus, as SAQ 1 of Block 5 implies, sulphur exists in oxidation numbers −2, 0, +2, +4 and +6; as usual, the highest oxidation number is equal to the Group number in Figure 92.

The so-called inert pair effect is most clearly seen among the elements of Periods 4–6 in Groups III–V (Figure 104): from Period 3 to Period 6, the oxidation number that is two less than the Group number tends to increase in stability relative to the Group number oxidation state.

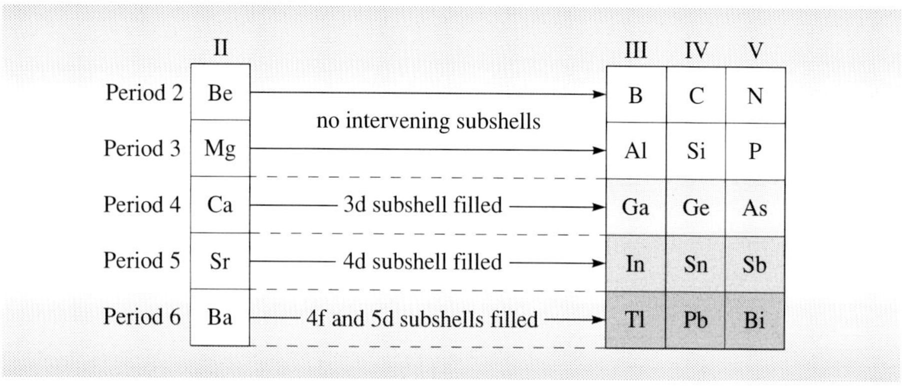

Figure 104 The Groups and elements in which the so-called inert pair effect is seen; the effect increases with the darkness of the shading.

■ Which elements in Figure 104 show the most marked inert pair effect, and which of their oxidation numbers does the effect stabilize?

□ Thallium(I), lead(II) and bismuth(III); the effect is most marked in Period 6, and the oxidation number stabilized is two less than the Group number.

Chemical evidence of the stability of these oxidation numbers was considered in Sections 3.3, 4.4 and 5.8; in each case, the higher, Group number oxidation states, thallium(III), lead(IV) and bismuth(V) are all powerful oxidizing agents. Bi_2O_5, for example, loses oxygen very easily and forms Bi_2O_3.

Figure 104 shows that the phenomenon occurs among those elements that immediately follow the filling of f and/or d subshells in the Periodic Table. It is most marked in Period 6, where the filling of the 4f *and* 5d subshells in the 24 preceding elements will be associated with a significant increase in nuclear charge. This increased nuclear charge will tend to bind the valence electrons in thallium, lead and bismuth more tightly and resist their removal for ion or bond formation in higher oxidation numbers. In particular, when the outer p electrons have been lost, the ions Tl^+, $Pb^{2+}$and Bi^{3+} are left with the configuration $[Xe]4f^{14}5d^{10}6s^2$. The unusually high nuclear charge resists removal of the outer $6s^2$ electrons for bond formation in higher oxidation numbers. It is because of this resistance of the outer s^2 pair for bonding that the phenomenon is called the *inert pair effect*.

ACTIVITY Now try the following SAQs. You should be able to answer them simply by referring to Figures 92 and 104. ■

SAQ 49 By referring to Figure 92 only, state which element of the following pairs has (a) the higher first ionization energy, (b) the higher electronegativity, (c) the larger covalent radius:

(i) aluminium and sulphur; (ii) oxygen and tellurium; (iii) chlorine and germanium.

SAQ 50 Consider the compounds IBr, $CaCl_2$ and $CaMg_2$. One is ionic, one is covalent and one is metallic. Identify which is which, match each compound to one of the three descriptions (i)–(iii) below, and state which one contains discrete molecules, with the given formula, in the solid state.

(i) A white solid that melts at 782 °C. It is a poor conductor of electricity in the solid state, but a good one when molten or when dissolved in water.

(ii) A brown–black solid that melts at 41 °C to give a liquid with low electrical conductivity.

(iii) A silvery-looking solid that melts at 720 °C. It is an excellent conductor of electricity in both solid and liquid states.

SAQ 51 Substance A is a typical element and is classified as a semi-metal. It forms two normal oxides, AO_2 and AO_3, and the fluorides AF_4, A_2F_{10} and AF_6. Identify A. What will be the formula of its hydride?

SAQ 52 Consider the three oxides PbO, BaO and SeO_3. One is acidic, one is amphoteric and one is basic. Identify which is which.

SAQ 53 Samples of the oxides GeO_2, SnO_2 and PbO_2 in crucibles are all placed in a sand bath at a temperature of 650 °C. One of the oxides decomposes. Which one is it, and what will be the decomposition reaction?

OBJECTIVES FOR BLOCK 7

Now that you have completed Block 7, you should be able to do the following things:

1 Recognize valid definitions of, and use in a correct context, the terms, concepts and principles in the following Table.

List of scientific terms, concepts and principles used in Block 7

Term	Page no.	Term	Page no.
activated charcoal	30	limiting nutrient	84
amorphous substance	52	Mond process	32
amphiboles	41	neutral oxide	93
anisotropic structure	29	nitrogen fixation	50
buckminsterfullerene	30	nitrosyl ligand	56
cage structure	68	orthorhombic sulphur	89
carbonyl compound	32	Ostwald process	58
combustion	92	oxoacid general formula	76
condensation of oxoacids	80	oxoacid nomenclature	77
contact process	101	Pauling's first rule	78
coordinative saturation	96	Pauling's second rule	79
cross-linked polymer	46	phosphonium salts	69
Czochralski process	37	plastic sulphur	90
dehydrating agent	71	polybasic acid	75
dihedral angle	90	polyphosphazenes	73
diradical	90	polycrystalline silicon	37
disilylene	47	polysulphides	95
dry ice	33	pyroxenes	41
drying agent	70	Raschig process	54
eutrophication	84	Rochow process	45
float-zone process	37	silanes	45
Frasch process	89	silicides	36
fullerenes	30	silicon wafers	37
greenhouse effect	35	silicones (polysiloxanes)	45
hydrophilic substance	83	silylenes	38
hydrophobic substance	83	Speier hydrosilation process	45
inert pair effect	25	sublimation	33
interstitial carbides	32	sulphanes	94
isomorphous replacement	40	thermal process	71
isostructural form	28	thioacids	100
isotropic structure	29	vulcanization process	90
lime water test	34	wet acid process	71

2 Apply skills and understanding summarized in the Objectives of Blocks 1–6 to the chemistry of the elements of Groups III–VI.

3 Recall important items of information about the Group III elements, notably those summarized in Sections 3.1.6 and 3.4, and combine them to gain new insights about the chemistry of those elements. (SAQs 3–9)

4 Recall important items of information about the Group IV elements, notably those summarized in Sections 4.2.3 and 4.5, and combine them to gain new insights about the chemistry of those elements. (SAQs 19 and 31)

5 Recall important items of information about the Group V elements, notably those summarized in Sections 5.4, 5.5.8 and 5.9, and combine them to gain new insights about the chemistry of those elements. (SAQs 31 and 48)

6 Predict the formulae of ortho oxoacids from coordination number and oxidation number considerations. (SAQs 24–30, 32, 42 and 43)

7 Use Pauling's first rule to predict (approximately) the dissociation constants for each dissociation step in polybasic acids. (SAQs 26–30, 42 and 43)

8 Use Pauling's second rule to predict (approximately) the first dissociation constant of an acid, from a consideration of the value of t in the general formula $AO_t(OH)_n$. (SAQs 26–30, 42 and 43)

9 Derive structural information for an unknown oxoacid, given the value of t in the general formula $AO_t(OH)_n$. (SAQs 29, 30, 42 and 43)

10 Predict the formulae of di-, tri-, and higher polymeric acids formed by condensation from the orthoacid. (SAQs 32 and 44)

11 Appreciate that condensation to high polymers is more likely for weak (that is, more highly hydroxylated) acids than for strong ones. (SAQ 32)

12 Recall important items of information about the Group VI elements, notably those summarized in Sections 6.2.3 and 6.5, and combine them to gain new insights about the chemistry of those elements. (SAQs 35 and 48)

13 Use your understanding of the trends in the Periodic Table summarized in Sections 2.3 and 7 to make predictions about the chemistry of the elements of Groups III–VI. (SAQs 1, 2, 47 and 49–53)

SAQ ANSWERS AND COMMENTS

SAQ 1 (*Objective 13*) In both forms of phosphorus, and in N≡N molecules, each atom forms three bonds to other atoms of the same sort. Thus, in all these cases each Group V atom is trivalent, and by exercizing this valency it acquires an octet of electrons. In nitrogen, however, the atoms express the trivalency in multiple bonds; in the phosphorus allotropes there are three separate single bonds. These preferences exemplify a characteristic difference between second- and third-row elements.

SAQ 2 (*Objective 13*) (i) The outer electronic configuration of the sulphur atom is $3s^2 3p^4$ (Structure **52**).

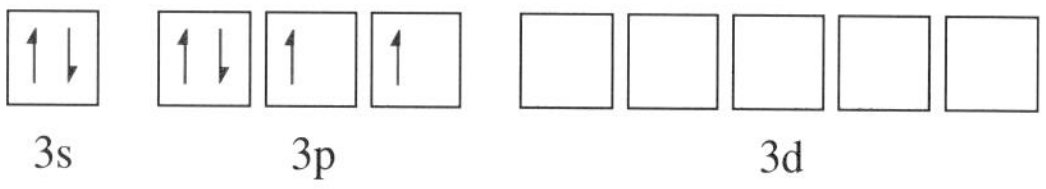

52

To form six S—F single electron-pair bonds, six unpaired electrons in six separate orbitals must be generated. This can be achieved by promoting an electron from each of the pairs in the 3s and 3p orbitals to separate 3d orbitals (Structure **53**).

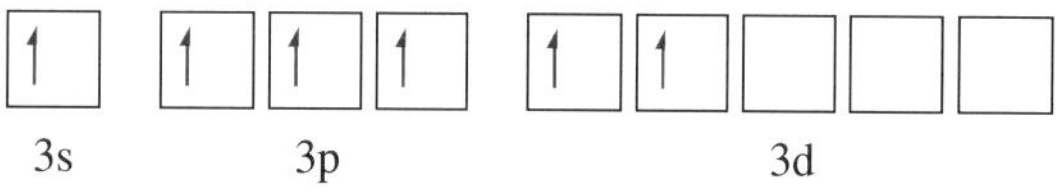

53

Thus, two 3d orbitals are involved in this bonding scheme.

(ii) Each of the six fluorine 2p orbitals involved in the bonding scheme contains one electron. When these six are added to the six sulphur electrons, the total is twelve. These twelve electrons exactly fill the four bonding orbitals and the two non-bonding ones. The total bond order is therefore four, distributed over six S—F bonds, which gives an average bond order of two-thirds.

SAQ 3 (*Objective 3*) The halides, BX_3, are stabilized by π-bonding in which non-bonding halogen electrons are delocalized into the $2p_z$ orbital on boron (Figure 10). In BH_3 this stabilization is impossible because hydrogen has no non-bonding electrons.

SAQ 4 (*Objective 3*) The short B(1)–B(3) distance is almost identical with that in B_2Cl_4 (Structure **9**), which suggests that this is a B—B single bond. The structure therefore contains one B—B single bond, six B—H terminal single bonds and four BHB three-centre bridging bonds. Each of these eleven bonds takes two valence electrons, making twenty-two in all. The four boron atoms have three valence electrons each, and the ten hydrogens each have one. Again, this comes to twenty-two; in other words, the bonds account for all of the valence electrons.

SAQ 5 (*Objective 3*) (a) The B_2O_3 structure is shown in Figure 105. For the sake of clarity it has been simplified by making the B—O—B sequences linear: this gives the structure the form of a planar sheet. However, the B—O—B angle is not 180°, and this destroys the sheet, making the structure a complex three-dimensional one. Nevertheless, Figure 105 shows correctly how the BO_3 triangles are joined. The oxygen atoms occur at the vertices, and each vertex is shared by two triangles. Therefore, each central boron is entitled to one half of each of its three surrounding oxygens. This gives the formula $BO_{1.5}$ or B_2O_3, as required.

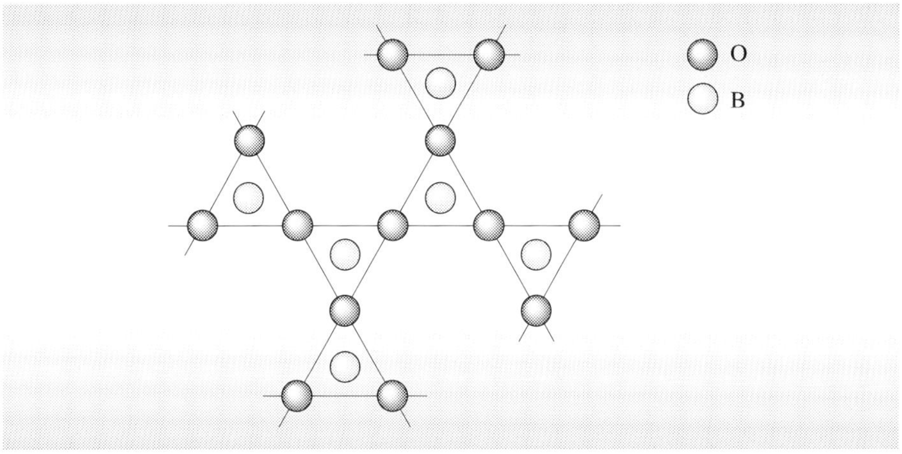

Figure 105 Simplified structure of B_2O_3.

(b) In the chain anion, only two of the three vertices in each triangle are shared with another triangle: the third is occupied by an unshared, terminal oxygen with a negative charge. Here then, each boron is assigned one half of each of two surrounding oxygens, and the whole of the third oxygen, along with its negative charge. The empirical formula of the anion is therefore BO_2^-, and the calcium salt of this anion must be $Ca(BO_2)_2$. Hence the proportion of the constituents of the mixture was 1 : 1.

SAQ 6 (*Objective 3*) We have argued that aluminium dissolves in alkali as well as acid because the protective oxide is amphoteric; that is, it is soluble in both media. As gallium oxide is amphoteric, and indium oxide is basic, this suggests that only gallium should dissolve in alkali. This is correct; the reaction corresponds to that of aluminium (Reaction 15):

$$2Ga(s) + 6H_2O(l) + 2OH^-(aq) = 2[Ga(OH)_4]^-(aq) + 3H_2(g) \quad (131)$$

SAQ 7 (*Objective 3*) We have noted the very marked tendency of aluminium to form the oxide, Al_2O_3. This suggests that the sulphate of aluminium should revert to the oxide at a lower temperature than does Na_2SO_4. This is correct: $Al_2(SO_4)_3$ decomposes at about 600 °C, whereas Na_2SO_4 remains undecomposed at 1 000 °C. Our conclusion can also be supported by lattice energy arguments: aluminium compounds tend to decompose more readily to the oxide because the small sizes and high charges of the Al^{3+} and O^{2-} ions give Al_2O_3 a large, negative lattice energy.

SAQ 8 (*Objective 3*) Two types of explanation have been used in the Course for such observations (Block 3, Section 13.3). Firstly, one can argue that the electronegativity difference between aluminium and the halogen decreases from AlF_3 to AlI_3, so covalent character will increase; secondly, in AlI_3, covalency is likely to be introduced by the polarization of the very large I^- ion by the small, highly charged Al^{3+} ion. This effect will diminish as the anion becomes smaller, and less polarizable, in the sequence I^-, Br^-, Cl^- and F^-.

SAQ 9 (*Objective 3*) Lead(II) and bismuth(III); we saw that the stability of the oxidation number that is two less than the Group number increases down Groups III–V. Thus, in Groups IV and V, lead and bismuth are the elements in which it should be most prominent. Note that lead(II) and bismuth(III) are the oxidation numbers that are associated with the Tl^+ electronic configuration, $[Xe]4f^{14}5d^{10}6s^2$, which contains an 'inert pair' of electrons in the 6s subshell. The conclusion is correct: lead(IV) and bismuth(V) are even less stable with respect to lead(II) and bismuth(III) than thallium(III) is with respect to thallium(I).

SAQ 10 (*Objective 2*) A metal complex, or coordination compound, consists of a metal atom or ion bonded to several neutral or anionic ligands. Carbon monoxide, CO, is a neutral unidentate ligand, which forms a four-coordinate complex with Ni; it coordinates through the carbon atom and the complex, $Ni(CO)_4$, is tetrahedral.

SAQ 11 (*Objective 2*) CO has a permanent dipole moment, so it has a rotational spectrum in the microwave (Block 6, Section 2.2). The dipole changes on stretching, and so CO will also absorb in the infrared (Block 6, Section 3.1).

SAQ 12 (*Objective 2*) The central carbon atom has four valence electrons, which together with two electrons from each oxygen, makes eight in all. These are distributed in two double bonds to the oxygens; the bond order for each bond is therefore 2. VSEPR theory predicts a linear shape for two repulsion axes (see Structure **16**, p. 33).

SAQ 13 (*Objective 2*) A linear molecule has $3n - 5$ normal modes. For CO_2 this will be four — two stretching modes and two degenerate bending modes. CO_2 is a linear molecule with $\mathbf{D}_{\infty h}$ symmetry, so it has a centre of symmetry and the mutual exclusion rule will apply. The symmetric stretch, ν_1, will be Raman active, and the bend, ν_2, and the antisymmetric stretch, ν_3, will be infrared active (see the discussion of linear triatomic molecules in Block 6, Section 4.3).

SAQ 14 (*Objective 2*) CO has a triple bond, whereas the carbon–oxygen bonds in CO_2 are double bonds. In the harmonic oscillator model

$$\nu = \frac{1}{2\pi}\sqrt{\left(\frac{k}{\mu}\right)} \tag{132}$$

We would expect the force constant, k, to be greater in CO, and therefore the C—O stretch should occur at a higher frequency than in CO_2. In practice the stretching frequency of CO is 2 140 cm^{-1} and the two stretching frequencies of CO_2 are found at 2 349 cm^{-1} (antisymmetric stretch, infrared) and 1 330 cm^{-1} (symmetric stretch, Raman); the large difference between the antisymmetric and symmetric stretch frequencies for CO_2 is due to the coupling of the two vibrations (Block 6, Section 4.6). The effect of increasing the isotope mass (and therefore μ) is to decrease the stretching frequency, so ν_1 and ν_2 for ^{13}CO should be at lower frequencies than the corresponding ^{12}CO vibrations.

SAQ 15 (*Objective 2*) According to the Ellingham diagram (see Block 1, Section 17), reduction of SiO_2 with C is not possible below 1 950 K.

$$\tfrac{1}{2}SiO_2(s) + C(s) = CO(g) + \tfrac{1}{2}Si(s) \tag{31}$$

$$\Delta H^{\ominus}_{m} = \Delta H^{\ominus}_{f}(CO, g) + \tfrac{1}{2}\Delta H^{\ominus}_{f}(Si, s) - \tfrac{1}{2}\Delta H^{\ominus}_{f}(SiO_2, s) - \Delta H^{\ominus}_{f}(C, s)$$

$$= \{-110.5 + 0 - (\tfrac{1}{2} \times -910.9) - 0\} = +344.9\,\text{kJ mol}^{-1}$$

$$\Delta S^{\ominus}_{m} = \Delta S^{\ominus}_{f}(CO, g) + \tfrac{1}{2}\Delta S^{\ominus}_{f}(Si, s) - \tfrac{1}{2}\Delta S^{\ominus}_{f}(SiO_2, s) - \Delta S^{\ominus}_{f}(C, s)$$

$$= \{197.7 + (\tfrac{1}{2} \times 18.8) - (\tfrac{1}{2} \times 41.8) - 5.7\} = +180.5\,\text{J K}^{-1}\,\text{mol}^{-1}$$

Using the relationship $\Delta G^{\ominus}_{m} = \Delta H^{\ominus}_{m} - T\Delta S^{\ominus}_{m}$ when $\Delta G^{\ominus}_{m} = 0$,

$$T = \frac{\Delta H^{\ominus}_{m}}{\Delta S^{\ominus}_{m}}$$

$$= \frac{344.9 \times 10^3}{180.5}\,\text{K}$$

$$= 1\,910.8\,\text{K}$$

The above calculation assumes that $\Delta H^{\ominus}_{m}$ and $\Delta S^{\ominus}_{m}$ do not change with T, and that none of the elements or compounds undergoes a phase change. In fact, silicon does melt within the range of our plot (m.t. 1 683 K); $\Delta H^{\ominus}_{fus}(Si)$, however, is small and there is very little change in slope at this point on the Ellingham diagram.

98% pure silicon is prepared commercially by the reduction of SiO_2 with carbon in an electric furnace, using a slight excess of SiO_2 to prevent formation of silicon carbide, SiC (via Equation 50, p.47).

SAQ 16 (*Objective 2*) SiF_5^- Si has four electrons and the five fluorines provide five electrons. Together with the single charge on the ion, there is a total of ten electrons, or five bonding pairs: these are used to make five single bonds to the fluorines, and so the shape of SiF_5^- expected from VSEPR theory is trigonal bipyramidal; its symmetry point group is $\mathbf{D}_{3h}$.

SiF_6^{2-} The six fluorine ligands together provide six electrons; the anion possesses a charge of −2, so there are twelve electrons in total, equivalent to six electron pairs. The six fluorines around the silicon form an octahedral structure, with the symmetry point group $\mathbf{O}_h$.

SAQ 17 (*Objective 2*)

$$CCl_4(l) + 2H_2O(l) = CO_2(g) + 4HCl(g) \tag{34}$$

$$\Delta G_m^{\ominus} = \Delta G_f^{\ominus}(CO_2, g) + 4\Delta G_f^{\ominus}(HCl, g) - \Delta G_f^{\ominus}(CCl_4, l) - 2\Delta G_f^{\ominus}(H_2O, l)$$

$$= \{-394.4 + (4 \times -95.3) - (-65.2) - (2 \times -237.1)\}\ \text{kJ mol}^{-1}$$

$$= -236.2\ \text{kJ mol}^{-1}$$

$$SiCl_4(l) + 2H_2O(l) = SiO_2(s) + 4HCl(g) \tag{35}$$

$$\Delta G_m^{\ominus} = \Delta G_f^{\ominus}(SiO_2, s) + 4\Delta G_f^{\ominus}(HCl, g) - \Delta G_f^{\ominus}(SiCl_4, l) - 2\Delta G_f^{\ominus}(H_2O, l)$$

$$= \{-856.6 + (4 \times -95.3) - (-619.8) - (2 \times -237.1)\}\ \text{kJ mol}^{-1}$$

$$= -143.8\ \text{kJ mol}^{-1}$$

So the hydrolysis of CCl_4 is thermodynamically *more* favourable than that of $SiCl_4$. (However, the equilibrium constant for the $SiCl_4$ hydrolysis is $1.1 \times 10^{25}\ \text{mol}^2\,\text{l}^{-2}$ at 298.15 K, so that can hardly be said to be unfavourable!) This result will be very surprising for those of you who have worked with silicon tetrachloride, since you will know that $SiCl_4$ is a fuming liquid that hydrolyses very easily, whereas carbon tetrachloride is a common solvent which is insoluble in water and does not react with it! Why is this? CCl_4 must be *kinetically stable* to reaction with water.

SAQ 18 (*Objective 2*) Using values from the *Data Book*:

(a) $$CH_4(g) + 2O_2(g) = CO_2(g) + 2H_2O(l) \tag{133}$$

$$\Delta G_m^{\ominus} = \Delta G_f^{\ominus}(CO_2, g) + 2\Delta G_f^{\ominus}(H_2O, l) - \Delta G_f^{\ominus}(CH_4, g) - 2\Delta G_f^{\ominus}(O_2, g)$$

$$= \{(-394.4) + (2 \times -237.1) - (-50.7) - 0\}$$

$$= -817.9\ \text{kJ mol}^{-1}$$

(b) $$C_2H_6(g) + \tfrac{7}{2}O_2(g) = 2CO_2(g) + 3H_2O(l) \tag{134}$$

$$\Delta G_m^{\ominus} = 2\Delta G_f^{\ominus}(CO_2, g) + 3\Delta G_f^{\ominus}(H_2O, l) - \Delta G_f^{\ominus}(C_2H_6, g) - \tfrac{7}{2}\Delta G_f^{\ominus}(O_2, g)$$

$$= \{(2 \times -394.4) + (3 \times -237.1) - (-32.8) - 0\}$$

$$= -1\,467.3\ \text{kJ mol}^{-1}$$

(c) $$SiH_4(g) + 2O_2(g) = SiO_2(s) + 2H_2O(l) \tag{135}$$

$$\Delta G_m^{\ominus} = \Delta G_f^{\ominus}(SiO_2, s) + 2\Delta G_f^{\ominus}(H_2O, l) - \Delta G_f^{\ominus}(SiH_4, g) - 2\Delta G_f^{\ominus}(O_2, g)$$

$$= \{(-856.6) + (2 \times -237.1) - 56.9 - 0\}$$

$$= -1\,387.7\ \text{kJ mol}^{-1}$$

(d) $$Si_2H_6(g) + \tfrac{7}{2}O_2(g) = 2SiO_2(s) + 3H_2O(l) \tag{136}$$

$$\Delta G_m^{\ominus} = 2\Delta G_f^{\ominus}(SiO_2, g) + 3\Delta G_f^{\ominus}(H_2O, l) - \Delta G_f^{\ominus}(Si_2H_6, g) - \tfrac{7}{2}\Delta G_f^{\ominus}(O_2, g)$$

$$= \{(2 \times -856.6) + (3 \times -237.1) - 127.3 - 0\}$$

$$= -2\,551.8\ \text{kJ mol}^{-1}$$

SAQ 19 (*Objective 4*) For carbon, no hexahalo species are expected because the maximum coordination number is four. F^- is the smallest halogen ion; the ionic radii increase through Cl^- and Br^- to reach a maximum at I^-. The maximum coordination number for silicon is six, but the *size* of the Group IV atom increases from silicon(IV) through germanium(IV) to tin(IV). Thus, we expect it to become increasingly easy to pack the larger halogens around the central atom down the Group. This is demonstrated by the occurrence of SiF_6^{2-}, $GeCl_6^{2-}$, $SnBr_6^{2-}$ and SnI_6^{2-}.

SAQ 20 (*Objective 4*) Because the Si—Si single bond is strong and the 2p–2p π overlap is not good, it is preferable for Si to form single bonds rather than double. The only way found so far of preventing Si=Si bonds from polymerizing, is to prevent their close approach by bulky ligands; hydrogen atoms would not be able to do this.

SAQ 21 (*Objective 2*) $\chi_{Si} = 1.8$ and $\chi_{Cl} = 3.0$. The ionic resonance energy calculated by Pauling's equation is thus

$$C(\chi_{Cl} - \chi_{Si})^2 = 96.5(3.0 - 1.8)^2 = 139\,\text{kJ mol}^{-1}$$

The bond enthalpy terms for Si—Si, Cl—Cl and Si—Cl are 226, 243 and 400 kJ mol^{-1}, respectively. The extra bond strength for Si—Cl is therefore measured as $400 - (226 + 243)/2 = 165.5\,\text{kJ mol}^{-1}$. The calculated value of the ionic resonance energy is therefore in reasonable agreement with the experimental value.

SAQ 22 (*Objective 2*) The HNO_2 molecule is shown in Structure **27**. It is planar, with no rotational axes in the plane of the molecule. It therefore belongs to the $\mathbf{C_s}$ symmetry point group.

SAQ 23 (*Objective 2*) The Lewis structure and structural formula of one resonance form of NO_3^- are shown in Structures **54** and **55**, respectively. The ion is planar with a C_3 axis. There are three C_2 axes perpendicular to the C_3 axis, and so the nitrate ion has $\mathbf{D_{3h}}$ symmetry.

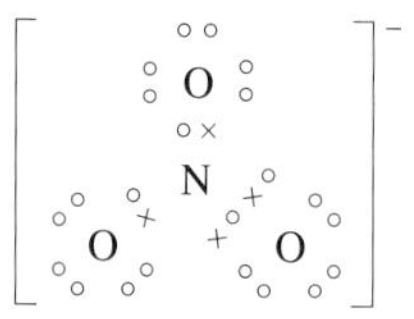

54

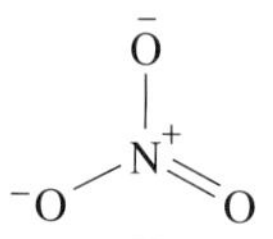

55

SAQ 24 (*Objective 6*) Assuming three-coordination for nitrogen(V), an oxoanion of stoichiometry NO_3 is formed, which bears a charge of $(+5 - 6) = -1$. The corresponding oxoacid thus requires only one proton for neutrality and its formula is HNO_3 (nitric acid).

For boron, the three-coordinate oxoanion has stoichiometry BO_3, which carries a resultant charge of $(+3 - 6) = -3$. The neutral acid needs three protons, so the formula of the ortho acid is H_3BO_3 (orthoboric acid).

SAQ 25 (*Objective 6*) Acids with $t = 0$ are, for example, H_4SiO_4, H_3BO_3 and HOCl; acids with $t = 1$ are, for example, H_3PO_4 and $HClO_2$; acids with $t = 2$ are, for example, H_2SO_4, HNO_3 and $HClO_3$. Finally, an acid with $t = 3$ is $HClO_4$.

SAQ 26 (*Objectives 6,7 and 8*) The chlorine oxoacids all have one —OH group, and zero, one, two and three terminal oxygens, respectively. The molecules are shown in Structures **56**–**59**.

HO — Cl

56

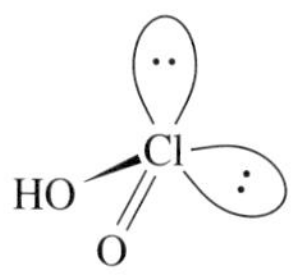

57

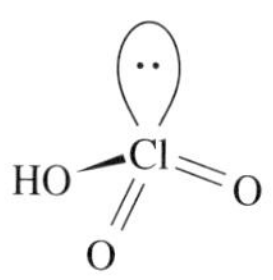

58

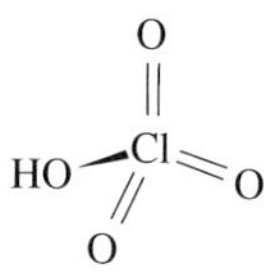

59

Chlorous acid, $HClO_2$, has four repulsion axes, two of which are lone pairs; it therefore adopts a bent shape. Chloric acid, $HClO_3$, has four repulsion axes, of which one is a lone pair; its shape is pyramidal. Perchloric acid, $HClO_4$, has four repulsion axes, with no lone pairs, and so we expect a structure based on the tetrahedral shape. According to Pauling's second rule, the acid strength is expected to increase in the direction

$$HClO < HClO_2 < HClO_3 < HClO_4$$

SAQ 27 (*Objectives 6, 7 and 8*) The formulae of the fourth-Period oxoacids are estimated from their oxoanion charge as follows:

GeO_4 charge (+4 – 8) = –4. Therefore the acid is H_4GeO_4, with $t = 0$, a weak acid. (In fact, the measured first dissociation constant, K_1, for germanic acid is 2.6×10^{-9} mol l^{-1}.)

AsO_4 charge (+5 – 8) = –3. Therefore the acid is H_3AsO_4, with $t = 1$, a moderately weak acid. (K_1 is measured as 5.6×10^{-2} mol l^{-1}.)

SeO_4 charge (+6 – 8) = –2. Therefore the acid is H_2SeO_4, with $t = 2$, a strong acid. (K_1 is not reported. K_2 is given as 1.2×10^{-2} mol l^{-1}; as Pauling's rules suggest, K_2 for this $t = 2$ acid is of the same order as K_1 for a $t = 1$ acid.)

BrO_4 charge (+7 – 8) = –1. Therefore the acid predicted is $HBrO_4$, with $t = 3$, a very strong acid. In fact, it is so strong that no reliable value for K_1 has been obtained.

SAQ 28 (*Objectives 6, 7 and 8*)

Using Pauling's second rule,

$t = 1$ for H_5IO_6
$t = 3$ for $HReO_4$
$t = 0$ for H_3AsO_3
$t = 2$ for H_2CrO_4

so the order of acid strength should be

$$HReO_4 > H_2CrO_4 > H_5IO_6 > H_3AsO_3$$

SAQ 29 (*Objectives 6, 7, 8 and 9*) The acid of stoichiometry H_3PO_3 (phosphorous acid) has a first dissociation constant typical of a moderately weak acid; that is, $t = 1$. According to Pauling's second rule, the dissociation constant for an acid with $t = 0$ would be expected to be around 10^{-8} mol l^{-1}, which suggests that H_3PO_3 *is not* the hydroxy acid $P(OH)_3$, but contains one P=O group. The correct structure for this compound is shown in Figure 106.

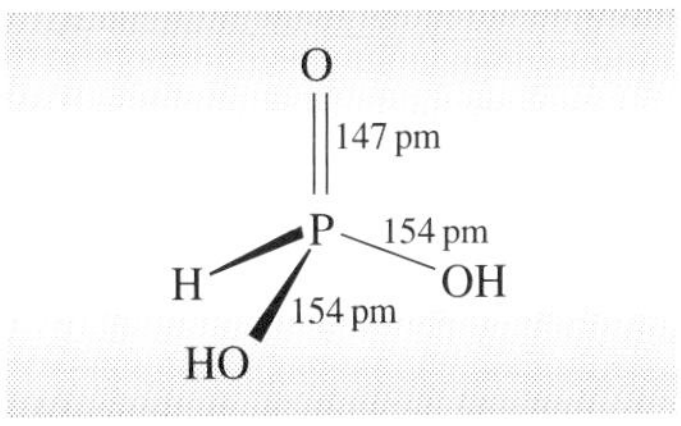

Figure 106 The structure of phosphorous acid, H_3PO_3 (showing the interatomic distances found by X-ray diffraction studies of the solid).

SAQ 30 (*Objectives 6, 7 8 and 9*) A K_1 value of 1×10^{-2} mol l^{-1} for H_3PO_2 (hypophosphorous acid) suggests that this is an acid with $t = 1$; that is, it probably contains one P=O group. This acid must accordingly contain two P—H groups and one P—OH. The likely structure is therefore the four-coordinate tetrahedral arrangement shown in Structure **60**.

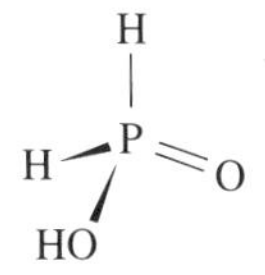

60

SAQ 31 (*Objectives 4 and 5*) Calcium precipitates as insoluble $CaCO_3$ when hard water is boiled, because the HCO_3^- present in solution is thereby converted into CO_3^{2-}:

$$Ca^{2+}(aq) + 2HCO_3^-(aq) = CaCO_3(s) + H_2O(l) + CO_2(g) \quad (137)$$

Calgon prevents $CaCO_3$ forming by making a stable soluble complex with calcium; that is, the calcium remains in solution.

SAQ 32 (*Objectives 6, 10 and 11*) From the formula for chromate ion, CrO_4^{2-}, we derive the molecular formula for chromic acid as H_2CrO_4. As you have discovered in SAQ 28, this is an acid with $t = 2$, that is, a strong acid. (In fact, $K_1 = 1.8 \times 10^{-1}\,\text{mol}\,\text{l}^{-1}$.)

Chromic acid is not expected to have a strong tendency to condense. In fact, you may remember from the Summer School experiment that the condensation equilibrium involves the hydrogen chromate ion, $HCrO_4^-$, which by Pauling's first rule is a much weaker acid ($K = 3.2 \times 10^{-7}\,\text{mol}\,\text{l}^{-1}$) than its parent H_2CrO_4:

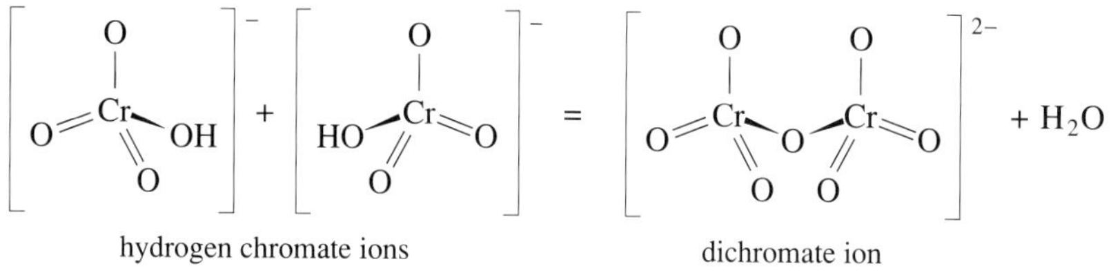

(138)

The parent acid of the dichromate ion is dichromic acid, $H_2Cr_2O_7$. It is a strong acid, which has no tendency to polymerize further. No metachromates are known to exist.

SAQ 33 (*Objective 2*) There are three different types of fluorine atoms in liquid SbF_5 (Figure 107). The bridging fluorines are labelled b, the fluorines *cis* to both bridges are labelled c, and those *trans* to one bridge and *cis* to another are labelled t.

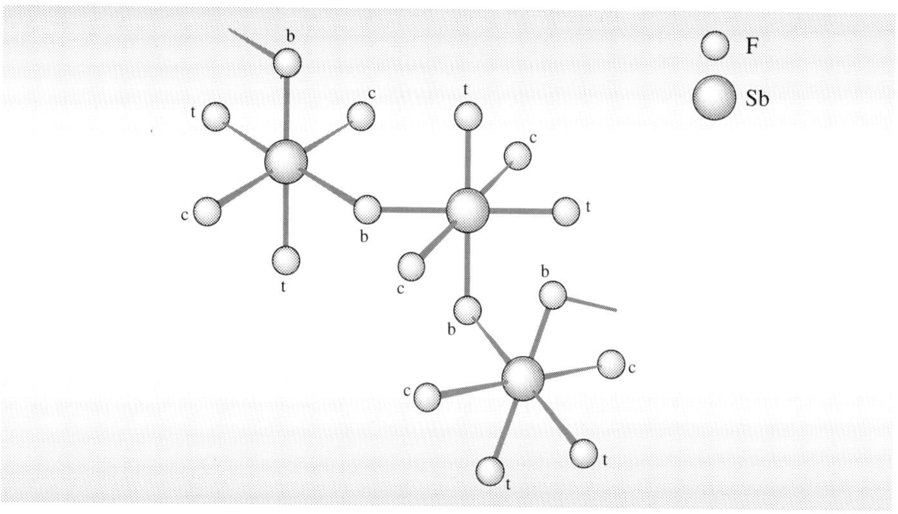

Figure 107 The structure of liquid SbF_5.

SAQ 34 (*Objective 2*) Paramagnetism cannot be explained on the basis of the Lewis structure (Figure 108a): there seems no reason why each S atom should not share two electrons with its neighbour to form a double bond. Using molecular orbital theory, we combine the 3s and 3p valence-shell electrons to obtain the partial molecular orbital energy-level diagram shown in Figure 108b.

Feeding the eight electrons from the 3p atomic orbitals into the molecular orbitals leaves the degenerate antibonding π orbitals half occupied by two electrons with parallel spin. Thus, S_2 is closely analogous to O_2 in its bonding and resulting paramagnetism.

SAQ 35 (*Objective 2*) The sum of the bond enthalpy terms of the products of Reaction 113 is

$$4\{B(\text{H}-\text{S})\} + 8\{B(\text{S}-\text{S})\}$$

The sum of the bond enthalpy terms for the reactants is also

$$4\{B(\text{H}-\text{S})\} + 8\{B(\text{S}-\text{S})\}$$

so the bond enthalpy approach predicts that Reaction 113 should be thermoneutral. In

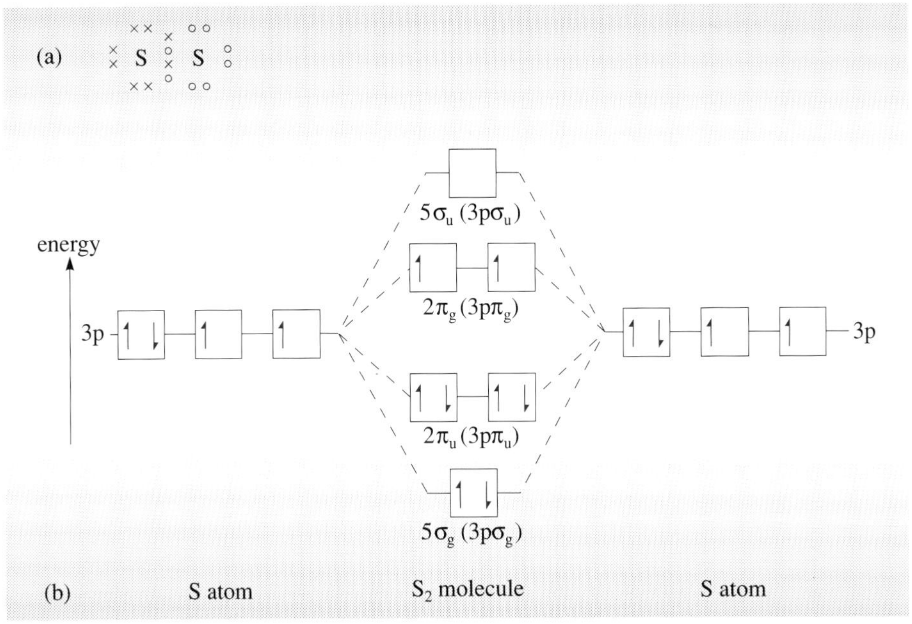

Figure 108 (a) Lewis structure of S_2; (b) partial molecular orbital energy-level diagram for S_2.

other words, because there is no change in the number of S—S or S—H bonds on forming a long-chain sulphane from H_2S and S_8, there should be no change in enthalpy.

As the entropy change in Reaction 113 is negative (a reduction in the total number of moles of gaseous species leads to a negative $\Delta S_m^\ominus$), this factor tends to favour the reverse (decomposition) reaction, particularly at temperatures close to and above room temperature. Thus, on balance, $\Delta G_m^\ominus$ for the reaction will be positive; that is, H_2S_5 is thermodynamically unstable.

SAQ 36 (*Objective 2*) For the reaction

$$H_2S(g) = H_2(g) + \tfrac{1}{2}S_2(g) \qquad (114)$$

$$\Delta S_m^\ominus = S^\ominus(H_2, g) + \tfrac{1}{2}S^\ominus(S_2, g) - S^\ominus(H_2S, g)$$
$$= (130.7 + 114.1 - 205.8)\,\text{J K}^{-1}\,\text{mol}^{-1}$$
$$= 39.0\,\text{J K}^{-1}\,\text{mol}^{-1}$$

$\Delta G_m^\ominus = 0$ at the decomposition temperature, so assuming $\Delta H_m^\ominus$ and $\Delta S_m^\ominus$ do not change with temperature,

$$\Delta H_m^\ominus = T\Delta S_m^\ominus$$

$\Delta H_m^\ominus$ for the decomposition is

$\Delta H_f^\ominus(H_2, g) + \tfrac{1}{2}\Delta H_f^\ominus(S_2, g) - \Delta H_f^\ominus(H_2S, g) = 0 + 64.2 - (-20.6) = 84.8\,\text{kJ mol}^{1}$

Hence the decomposition temperature, T

$$= \frac{\Delta H_m^\ominus}{\Delta S_m^\ominus} = \frac{84\,800\,\text{K}}{39.0}$$

$$= 2\,174\,\text{K}\ (1\,901\,^\circ\text{C})$$

In a multistage process (as for H_2O; see Case Study 1) it should be possible to reduce the decomposition temperature appreciably. A process such as:

$$H_2S(g) + CO(g) = H_2(g) + COS(g) \qquad (139)$$

$$4COS(g) + 2SO_2(g) = 4CO_2(g) + 3S_2(g) \qquad (140)$$

combined with the use of a suitable semipermeable membrane, which allows the passage of only the smallest molecule, would permit continuous withdrawal of hydrogen at temperatures as low as 1 000 K.

SAQ 37 (*Objective 2*) (a) and (b) The $\mathbf{D_{\infty h}}$ and $\mathbf{C_{2h}}$ forms of S_2Cl_2 are centrosymmetric; modes that do not disturb the symmetry of centrosymmetric molecules are not active in the infrared because they do not change the dipole moment (Block 6, Section 4.6). The S—S stretching mode will thus be observed in the Raman spectrum only. Of the two S—Cl stretching modes, only the antisymmetric stretch (S—Cl) will be observed in the infrared. Neither *cis*-planar, $\mathbf{C_{2v}}$, nor non-planar, $\mathbf{C_2}$, forms are centrosymmetric, so the arguments of Block 6 must be used to determine the symmetry character of the bond extensions.

If Δr_3 is taken to represent the S—S bond extension, and the pair Δr_1, Δr_2 to represent the two S—Cl extensions, we can see how these extensions transform under the symmetry operations of the $\mathbf{C_{2v}}$ point group. The results are set out in Table 24.

Table 24 Representation of the bond extensions for the $\mathbf{C_{2v}}$ structure of S_2Cl_2

C_{2v}	$\hat{I}$	$\hat{C}_2$	$\hat{\sigma}_v(xz)$	$\hat{\sigma}_v(yz)$
Δr_3	1	1	1	1
$\Delta r_1, \Delta r_2$	2	0	0	2

The representation for Δr_3, when compared with the $\mathbf{C_{2v}}$ character table in the *Data Book* corresponds to the irreducible representation for a_1 character. ν(S—S) thus has a_1 character; inspection of the final column of the $\mathbf{C_{2v}}$ table in the *Data Book* shows it to be active in both the infrared and the Raman.

On the other hand, the representation for Δr_1, Δr_2 is reducible, corresponding to the sum of a_1 and b_2 representations. This indicates that one of the ν(S—Cl) modes has a_1 character and the other has b_2 character. As the final column of the $\mathbf{C_{2v}}$ character table indicates, both modes are active in both the infrared and the Raman.

Under $\mathbf{C_2}$ symmetry the bond extensions transform as in Table 25. r_3 has 'a' symmetry, and the Δr_1, Δr_2 pair correspond to combination of a + b representations. The final column of the $\mathbf{C_2}$ character table shows that all $\mathbf{C_2}$ modes are active in both the i.r. and the Raman, so once more all three stretching modes appear both in the infrared and the Raman. Table 26 summarizes these conclusions.

Table 25 Representations of the bond extensions for the $\mathbf{C_2}$ structure of S_2Cl_2

C_2	$\hat{I}$	$\hat{C}_2$
Δr_3	1	1
$\Delta r_1, \Delta r_2$	2	0

Table 26 Infrared and Raman activity of vibrational modes of S_2Cl_2*

Symmetry	Activity of S—Cl stretch	Number of i.r.-active modes	Number of Raman-active modes
$\mathbf{C_{2v}}$	i.r. and Raman	2	2 (1 polarized)
$\mathbf{C_2}$	i.r. and Raman	2	2 (1 polarized)
$\mathbf{D_{\infty h}}$	i.r. and Raman	1	2 (1 polarized)

* The analysis of the infrared and Raman activity of the $\mathbf{C_{2h}}$ form is outside the scope of this Course.

(c) On the basis of the above analysis of stretching modes, it is quite easy to distinguish the centrosymmetric structural possibilities ($\mathbf{D_{\infty h}}$ and $\mathbf{C_{2h}}$) from the others.

However, the *cis*-planar $\mathbf{C_{2v}}$ geometry cannot be distinguished from the non-planar $\mathbf{C_2}$ form on the basis of stretching modes alone. (It should, in fact, be possible to distinguish between these on the basis of a full vibrational analysis, since one of the bending modes in $\mathbf{C_{2v}}$ symmetry has a_2 character and is active in the Raman only.)

In fact, S_2Cl_2 (and indeed also S_2Br_2 and one isomer of S_2F_2) has the non-planar $\mathbf{C_2}$ geometry. Dimensions of S_2Cl_2 are as follows:

r(S—Cl)/pm	r(S—S)/pm	$\widehat{SSCl}$	dihedral angle
207	197	107°	82.5°

As in the case of the element and the sulphanes, molecular geometry is determined by the dihedral angle.

SAQ 38 (*Objective 2*) The central sulphur atom in $S{=}SX_2$ has three repulsion axes involving bonding electron pairs, one double and two single, which account for four of its six valence electrons. The non-bonding pair must form a fourth repulsion axis, leaving the molecule with a pyramidal structure (Figure 109). The symmetry, accordingly, is $\mathbf{C}_s$.

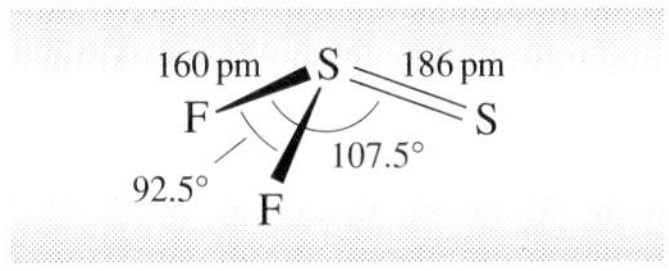

Figure 109 Geometry of the S_2F_2 molecule.

Applying the arguments of Block 6 to S_2F_2 in $\mathbf{C}_s$ symmetry, taking Δr_1 to represent the S=S bond extension and Δr_2, Δr_3 to represent the two S—X bond extensions, it is clear that Δr_1 has a′ symmetry, and the Δr_2, Δr_3 pair has a representation corresponding to a combination of a′ and a″ (Table 27).

Table 27 Representations of the bond extensions of the $\mathbf{C}_s$ structure of S_2X_2

C_s	$\hat{I}$	σ
Δr_1	1	1
Δr_2, Δr_3	2	0

The $\mathbf{C}_s$ character table shows that both a′ and a″ modes are simultaneously active in the infrared and Raman. Thus, all three stretching modes should be observed in the infrared, as was the case for the $\mathbf{C}_{2v}$ and $\mathbf{C}_2$ forms of S_2X_2 discussed in SAQ 37. The *number* of infrared stretching modes observed is no help in distinguishing $\mathbf{C}_s$ from $\mathbf{C}_{2v}$ or C_2.

However, the group frequency approach is helpful here. The mode corresponding to S=S stretching is expected at higher wavenumber than that corresponding to vibration of the *single* S—S bond in the $\mathbf{C}_{2v}$ and $\mathbf{C}_2$ structural possibilities for the XSSX linkage. In addition, the S=S stretch should appear more strongly in the Raman spectrum than an S—S stretch. This is because intensity of a Raman band depends on polarizability, and a double bond is more polarizable than a single bond.

S_2F_2 exists as two isomers; one is the $\mathbf{C}_s$, form described in Figure 109, and the other is a $\mathbf{C}_2$ form analogous to S_2Cl_2.

SAQ 39 (*Objective 2*) Sulphur(IV) in SO_2 is not making use of all its electrons in bonding to oxygen. The non-bonding pair forms a third repulsion axis, so the molecule is bent (Structure **43**). Sulphur(VI) in SO_3 is using all its valence electrons in bonding to oxygen. There are no non-bonding pairs, so the molecule is trigonal planar (Structure **44**).

SO_2 is polar (has a net dipole moment) but SO_3 is not. Thus, SO_2 fulfils the prerequisite for showing a rotational spectrum, so the S=O bond length in SO_2 can be found from an examination of its microwave spectrum; SO_3, however, does not satisfy the general selection rule for giving a rotational spectrum. It will absorb infrared radiation, though, and a (less accurate) value for the sulphur–oxygen internuclear distance can be obtained from an analysis of the rotational fine structure of the vibrational spectrum of SO_3.

SAQ 40 (*Objective 2*) (a) The couple of interest is $E^\ominus(SO_4^{2-}|SO_2)$, which corresponds to the following reaction:

$$SO_4^{2-}(aq) + 4H^+(aq) + 2e^- = SO_2(g) + 2H_2O(l) \qquad (141)$$

$E^\ominus$ for Equation 141 can be obtained by subtraction:

$$E^\ominus(SO_4^{2-}|S) - E^\ominus(SO_2|S) = 0.352 - 0.451 = -0.099\ V$$

(b) $E^\ominus$ for the oxygen couple is larger than that for the $SO_4^{2-}|SO_2$ couple, and therefore the atmospheric oxidation reaction is thermodymamically possible:

$$\tfrac{1}{2}O_2(g) + H_2O(l) + SO_2(g) = SO_4^{2-}(aq) + 2H^+(aq) \qquad (142)$$

SAQ 41 (*Objective 2*) According to Table 21, $E^\ominus(S|H_2S)$ is less positive than $E^\ominus(SO_2|S)$, so H_2S will reduce SO_2 to sulphur. The overall equation for the reaction is:

$$SO_2(g) + 2H_2S(g) = 3S(s) + 2H_2O(l) \qquad (143)$$

This reaction is in fact a very useful one; it is used in the Claus process for the removal of H_2S from natural gas and other petroleum products. Some H_2S is first burnt to produce SO_2, which is then passed through a stream of H_2S. The sulphur produced can be sold at a profit.

SAQ 42 (*Objectives 6, 7, 8 and 9*) Sulphur has six valence electrons and so can form six covalent bonds. We also know that elements of the third Period prefer four coordination in their oxoanions. Sulphur(VI) coordinates four O^{2-} ions to form a complex of stoichiometry SO_4, with resultant charge (+6 − 8) = −2, so the oxoacid is sulphuric acid, H_2SO_4 (Structure **61**). The formation of disulphuric acid, $H_2S_2O_7$, from two molecules of sulphuric acid is represented in Reaction 144:

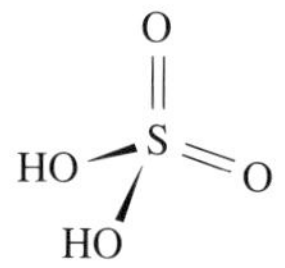

61

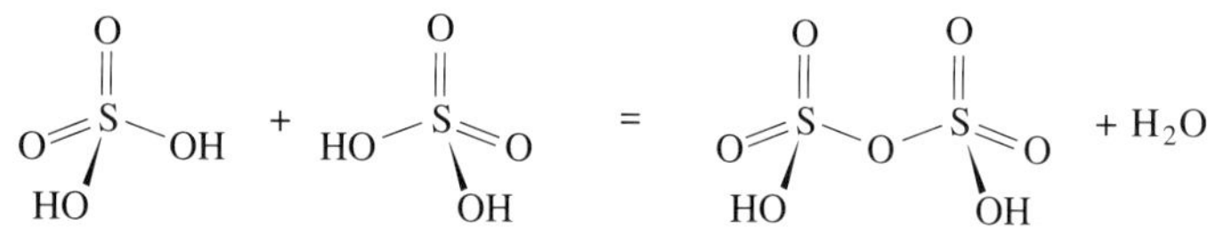

(144)

SAQ 43 (*Objectives 6, 7, 8 and 9*) From Pauling's second rule we can say that, as H_2SO_4 has two terminal oxygen atoms (t = 2), it should be a strong acid.

Applying Pauling's first rule,

$$\frac{K_1}{K_2} \sim 10^5, \text{ so } K_2 = 1 \times 10^3 \div 10^5 = 1 \times 10^{-2}\,\text{mol}\,\text{l}^{-1}$$

The measured value is $1.2 \times 10^{-2}\,\text{mol}\,\text{l}^{-1}$.

SAQ 44 (*Objective 2*) The Raman spectrum of the aqueous solution clearly shows an absorption at *c*. 2 600 cm^{-1}, which is attributable to ν(S—H) stretching, but there is no sign of ν(O—H) in the 3 500–3 000 cm^{-1} region. This suggests that only one form is present to any extent. (You should note that the spectrum was obtained in water, which is an extremely weak Raman scatterer, and rarely obscures any ν(O—H) due to solute.) In the solid state there is no peak in the Raman in the 3 200 cm^{-1} region, in contrast to the solid-state spectrum of hydrogen sulphate ion, HSO_4^-, where O—H stretching modes are observed at 3 200 cm^{-1}.

SAQ 45 (*Objective 2*) S_4N_4 decomposes into its elements, N_2 and S_8 as follows:

$$2S_4N_4(s) = S_8(s) + 4N_2(g); \qquad \Delta H_m^{\ominus} \approx -460\,\text{kJ}\,\text{mol}^{-1} \qquad (145)$$

The stability of the S_8 and N_2 molecules provides the driving force for this reaction.

SAQ 46 (*Objective 2*) Possible resonance structures which describe the bonding in S_2N_2 are shown in Figure 110.

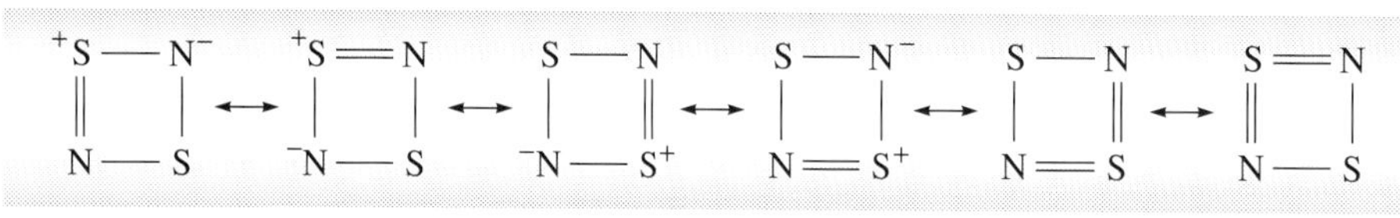

Figure 110 Resonance forms of S_2N_2.

SAQ 47 (*Objective 13*) SO_2 is a non-linear gaseous molecule, with double bonds to oxygen and a coordination number of two (Structure **43**). We find that selenium dioxide, SeO_2, is a chain polymer containing the linkage shown in Structure **62**. There is less double bonding and the coordination number has increased to three. Tellurium dioxide contains four-coordinate TeO_4 units singly bonded into layer and three-dimensional structures. In the oxide of the metal polonium we find that PoO_2 adopts the ionic fluorite structure, and the coordination number increases to eight. Summarizing, we find a move away from covalent bonding towards ionic from the top to the bottom of the Group, in line with the increasing metallic character of the elements and at the same time we also find an increase in coordination number.

43

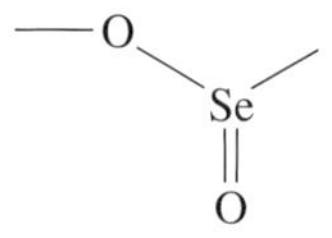

62

SAQ 48 (*Objectives 5 and 12*) We are told that NS is not stable and we know from Section 5.3.3 that NO is thermodynamically unstable with respect to its elements (Equation 78). However, in the case of NO the decomposition is undetectably slow at room temperature and is only observed at about 1 000–1 200 °C. Thus, the existence of NO at room temperature is due to kinetic factors.

SAQ 49 (*Objective 13*) (i) Aluminium precedes sulphur in Period 3 (Figure 92). Consequently, sulphur has the higher first ionization energy and electronegativity because these quantities show an overall increase across a Period; aluminium has the larger covalent radius because this quantity shows an overall decrease across a Period.

(ii) Oxygen lies above tellurium in Group VI (Figure 92). Consequently, oxygen has the higher first ionization energy and electronegativity, but tellurium has the larger covalent radius.

(iii) Germanium is in a later Period than chlorine, but chlorine is in a later Group (Figure 92). Consequently, the two effects involved in parts (i) and (ii) reinforce each other: chlorine has the higher first ionization energy and electronegativity; germanium has the larger covalent radius.

SAQ 50 (*Objective 13*) (i) $CaCl_2$; (ii) IBr; (iii) $CaMg_2$. Solid IBr contains discrete molecules.

The properties listed are those of (i) an ionic substance, (ii) a covalent substance, and (iii) a metallic substance. $CaCl_2$ is a combination of elements from Groups II and VII, so the electronegativity difference will be large, and the compound will therefore be ionic. IBr will be covalent because it is a combination of two elements of high electronegativity from Group VII. $CaMg_2$ will be a metallic alloy because it is a combination of metallic elements of low electronegativity from Group II.

SAQ 51 (*Objective 13*) A is tellurium and its hydride is TeH_2. The existence of the compounds AO_3 and AF_6 suggests that the highest valency and oxidation number of A is six. This implies a Group VI element. According to Figure 92, the semi-metal in Group VI is tellurium. The Group VI elements form hydrides of formula H_nX, where $n = 8 - 6$.

SAQ 52 (*Objective 13*) SeO_3 is acidic, PbO is amphoteric and BaO is basic. The acidity of oxides tends to increase across a Period and decrease down a Group. In Figure 92 selenium lies above and to the right of both barium and lead, so on both counts its oxide should be the most acidic. As lead is later than barium in Period 6, its oxide will be the less basic. Thus, BaO will be the basic oxide and PbO will be the amphoteric one.

SAQ 53 (*Objective 13*) PbO_2 is the oxide that decomposes. All three elements fall in the region of Figure 92 where the inert pair effect is strongly apparent in chemical behaviour. The effect causes the oxidation state two less than the Group number to increase in stability from the top to the bottom of a Group. Thus, the oxidation state of +4 corresponding to the Group number will decrease in stability from germanium to lead. PbO_2 is therefore the least stable dioxide, and on heating will be the first to decompose to an oxide in which lead has an oxidation number of +2:

$$2PbO_2(s) = 2PbO(s) + O_2(g) \qquad (146)$$

ACKNOWLEDGEMENTS

Grateful acknowledgement is made to the following sources for permission to reproduce material in this Block:

Text

McGough, R., 'Carbon', 'Nitrogen', 'Oxygen', and 'Sulphur', *Defying Gravity*, Viking, reprinted by permission of the Peters, Fraser and Dunlop Group Ltd.

Figures

Figure 13 Chemie in unserer Zeit, **20** (1986); *Figure 24* P. Stokes (1988), 'Where green hair is all the rage', *Daily Telegraph*, 14 July 1988; *Figure 29a* Richard Whittington-Egan; *Figure 29b* Topham Picture Source; *Figure 76* painting by George Richmond at the Royal Institution; *Figure 82* Mansell Collection; *Plates 1, 2, 7 and 12* Charles D. Winter from J. C. Kotz and K. F. Purcell, *Chemistry and Chemical Reactivity*, Saunders (2nd edn 1991); *Plates 3, 4, 6, 10 and 11* Natural History Museum, London; *Plate 5* Tony Waltham, Nottingham Trent University; *Plate 8* Johnson Matthey PLC.